Agnes Mary Clerke and the Rise of Astrophysics

Born in Ireland in the mid nineteenth century, Agnes Clerke achieved fame as the author of *A History of Astronomy during the Nineteenth Century*. Through her quarter-century career, she became the leading commentator on astronomy and astrophysics in the English-speaking world. Her rise to this status coincided with the spectacular rise of the 'new astronomy' and the foundation of the great American observatories from which she derived much of her information and inspiration.

This fascinating biography of Agnes Clerke describes the life and work of this extraordinarily erudite but unassuming woman. Her story is also a chronicle of the development of astronomy in the last decades of pre-Einstein science, and introduces many of the great figures in astronomy of that age, Huggins, Lockyer, Holden, Pickering, Gill, Vogel and Hale, and their achievements and their rivalries. Her life was closely linked with that of her talented journalist sister Ellen, whose activities throw further light on Agnes Clerke's personality and motivation. The story follows her friendship with William and Margaret Huggins, and her prolific correspondence with eminent astronomers of the time, such as David Gill of the Cape and George Ellery Hale of California. This biography of Agnes Clerke will fascinate not only scientists but everyone who admires intellectual achievement brought about through love of learning and sheer hard work.

Mary Brück gained her PhD in astronomy from the University of Edinburgh, where she went on to become a senior lecturer in astronomy. Her main research interest was in photographic stellar photometry and spectroscopy. Now retired, she has a special interest in nineteenth-century British and Irish women in astronomy, about whom she has written numerous articles. In 2001, Dr Brück was awarded the Lorimer Medal of the Astronomical Society of Edinburgh.

Agnes Mary Clerke.

Agnes Mary Clerke
and the Rise of Astrophysics

M. T. Brück

CAMBRIDGE
UNIVERSITY PRESS

CAMBRIDGE UNIVERSITY PRESS
Cambridge, New York, Melbourne, Madrid, Cape Town, Singapore, São Paulo

Cambridge University Press
The Edinburgh Building, Cambridge CB2 8RU, UK

Published in the United States of America by Cambridge University Press, New York

www.cambridge.org
Information on this title: www.cambridge.org/9780521808446

First published 2002
This digitally printed version 2008

A catalogue record for this publication is available from the British Library

Library of Congress Cataloguing in Publication data

Brück, M. T. (Mary T.)
  Agnes Mary Clerke and the rise of astrophysics / M. T. Brück
    p. cm.
  Includes bibliographical references and index.
  ISBN 0 521 80844 8
  1. Clerke, Agnes M. (Agnes Mary), 1842–1907. 2. Astronomers –
Ireland – Biography. 3. Astronomy – History – 19th century. I. Title.

QB36.C57 B78 2002
520′.92 – dc21
[B]   2001043451

ISBN 978-0-521-80844-6 hardback
ISBN 978-0-521-05579-6 paperback

For Emma Rose and Lily May

Contents

Acknowledgements

Agnes Clerke's career as a writer emerges most clearly from her substantial correspondence, and for that reason I owe a particular debt of gratitude to the archivists and librarians who have generously given me access to these illuminating materials, as noted in the list of references. In thanking them, I would like to make special mention of Mr Adam J. Perkins, Royal Greenwich Observatory archivist at Cambridge University Library; Mrs Dorothy Schaumberg, curator of the Mary Lea Shane archives of Lick Observatory; Dr Owen Gingerich, Harvard College Observatory; Mr D.W. Evans of the University Library, University of Exeter; Dr Jennifer Stine and Ms Shelley Irwin, Institute Archives, California Institute of Technology; Herr Michael Stanske, University Library, Heidelberg; Sister Ursula Clarke, Ursuline Convent, Cork; and also the archivists of the Library of Congress, Washington, the Memorial Library, University of Wisconsin, Madison and the Royal Institution, London. I also thank Mr John Wilson for transcripts from the Wilson Daramona diaries; Dr Ian Glass and Mrs Ethleen Lastovica for Cape photographs; and Dr W. Sheehan for references. I owe more than I can express to the librarians of our excellent Edinburgh libraries, especially Mr Angus Macdonald and Ms Karen Moran at the Royal Observatory Edinburgh, for ever-helpful assistance with the literature.

Negative results must also be recorded. I therefore acknowledge with thanks the trouble of searches made by archivists whose collections do not, in fact, contain relevant material. These include the Institut d'Astrophysique, Paris; Potsdam Astrophysical Observatory; Dunsink Observatory, Ireland; the Science Museum, London; the US Military Academy, West Point; Stonyhurst College, Lancashire; St Patrick's College, Maynooth; the Mill Hill Fathers, London; and the

Westminster Diocesan archives (the last were lost in the bombing of London during the Second World War).
I am grateful to Professor Michael J. Crowe and Dr Michael A. Hoskin for some much appreciated reassurance in their special fields. I have also been much encouraged by my friends and fellow enthusiasts for nineteenth-century astronomy, Dr Barbara Becker, Dr Allan Chapman and Dr Ian Elliott. As regards Agnes Clerke's earlier life in Ireland, I give my very special thanks for information generously given on her family and background to Mr Rickard Deasy, descendant of Agnes Clerke's first cousin and the family's closest relative; to the late Father James Coombes, who started me off on the trail of the Clerke and Deasy families, and to Mr Pat Cleary of Skibbereen for sharing his knowledge of the history and geography of his lovely town. Dr Owen Dudley-Edwards kindly provided wise guidance on the historical background. I would also like to mention members of my own family who took a practical interest in my project, especially Peter and Joan Brück, and Maev Conway-Piskorski. Finally, my warmest thanks go to Mrs Sheila O'Kelly Deasy, who made it an extra pleasure for me to peruse the Clerke books while I enjoyed the hospitality of her beautiful home in Co. Tipperary.

Acknowledgement of sources of illustrations

Portraits of Agnes and Ellen Clerke: Royal Astronomical Society (from Lady Huggins' *Appreciation*); The Clerke parents: R.H. Deasy; E.S. Holden: Mary Lea Shane archives of Lick Observatory, University of California; J.N. Lockyer: Royal Astronomical Society; George Ellery Hale: Hale archives, California Institute of Technology; Cape Observatory photographs: South African Astronomical Observatory; Skibbereen photographs: P.M. Brück. Spectrum of emission-line star: Royal Observatory Edinburgh. Starfields from Agnes Clerke's books have been reproduced by Media Services, UK Astronomy Technology Centre, Royal Observatory Edinburgh.

Introduction

The second half of the nineteenth century was a period of extraordinary scientific progress. In astronomy, bigger and better telescopes, and the techniques of spectroscopy and photography, brought about a revolution in humankind's vision of the universe. The documentation – in the English-speaking world at least – of the astronomical labours of that important era was almost entirely due to Agnes Mary Clerke, historian of astronomy and painstaking chronicler of astrophysical discovery as she witnessed it over thirty years of her active life. This remarkable woman, educated solely within her own family and through her own private studies, not only kept abreast of astronomical progress world-wide but also had a genuine understanding of the matters on which she reported and the gift of communicating them through her fluent and prolific writings. Her books – in particular her *Popular History of Astronomy during the Nineteenth Century*, first published in 1885 and reprinted over almost twenty years – are treasured by historians and by amateur lovers of astronomy alike as sources of reliable and enjoyable information on that period. She was also much in demand in her lifetime as a contributor to literary journals and encyclopaedias.

Agnes Clerke numbered among her circle of friends and correspondents many of the eminent astronomers of the day. Unobtrusive and gentle by nature, she nevertheless became an authority on astrophysics and a figure of respect in the literary world. This status was achieved, as is revealed in letters preserved in the archives of observatories in South Africa, the United States and Europe, through intense application and tireless enquiries from the experts. Agnes Clerke's erudition stretched beyond science; she was a classical scholar, a gifted linguist, an accomplished musician, and a person of deep religious conviction. Her story, from her upbringing in a remote town in Ireland in the dark days of the potato famine, to her thirty years residence in London, is linked with that of her sister Ellen who, in another sphere, was a minor poet and a journalist of some influence in the English Catholic press.

1 Family background in County Cork

The Clerkes

Agnes Mary Clerke, born on 10 February 1842 in Skibbereen, Co. Cork, was the second of three children of John William Clerke, manager of the Provincial Bank in that town, and his wife Catherine Mary née Deasy (Figure 1.1).[1]

The Clerkes were a well-known and extensive family in West Cork. According to one account, the founder of the line was a major in King William's army who stayed on in Ireland after 1691. In the nineteenth century the Clerke family records yield a remarkable number of highly talented persons. Agnes' grandfather, St John Clerke, was a much loved physician in Skibbereen; his cousin was the renowned Dr Jonathon Clerke of Bandon. The latter's son, Major Sir Thomas Henry Shadwell Clerke,[2] was made a knight of the Royal Hanoverian Order after service in the Peninsular War, and became a military journalist. He took a keen interest in the sciences and in 1823 became a founder member of the Royal Astronomical Society, to which Agnes Clerke was to be elected some 90 years later. He was also elected a Fellow of the Royal Society and of the Royal Geographical Society. He was made Foreign Secretary of the latter society on account of his linguistic prowess. Among other noted members of the Clerke clan in the nineteenth century was Skibbereen-born Thomas W. Clerke, LlD, Judge of the Supreme Court in the USA, author of important treatises on law and co-founder in 1841 of the Irish Emigrant Society.[3]

In the immediate family, Agnes Clerke's grandfather St John Clerke, the local medical doctor, was esteemed not only professionally but also for his energetic support of just causes. He and his brother Thomas, a corn miller, were prime movers in organising relief in the

Figure 1.1 Catherine and John Clerke, Agnes Clerke's parents.

form of food and employment for the victims of the near-famines of 1817 and 1822, as well as personally donating several hundred pounds 'to our starving poor'. A Protestant himself, Dr Clerke was equally strenuous in campaigning to secure legal rights for his Catholic fellow citizens in the days before Catholic emancipation (1829). To show his solidarity with them, he attended in person a general meeting of Catholics in the Skibbereen area, convened in November 1824 to protest at sectarian injustices and to support Catholic Rent, a fund initiated by the great Irish politician Daniel O'Connell to fight the cause by legal means.[4]

It was into this exemplary and liberal-minded family that Agnes Clerke's father, John William, was born in 1814, one of three brothers and a sister. All three brothers were educated at Trinity College Dublin. John, a scholar in classics also studied mathematics and the sciences, especially astronomy: his surviving annotated textbook, Robert Woodhouse's *Treatise on Astronomy*, testifies to his industry in this regard. On graduating he returned to his native town and, having worked briefly in a locally owned bank, was appointed manager of the London-based Provincial Bank (now the Allied Irish Bank) when it

opened its doors in Skibbereen in 1839. The bank, on Bridgetown or
Bridge Street, the town's main street, was the largest house in the street
and one of the most substantial in the entire town.

Here, on his appointment, John Clerke set up home with his
bride, Catherine Deasy, a sister of his friend and contemporary at
Trinity College, Rickard Morgan Deasy.

The Deasys

The Deasys belonged to an old Irish family with a romantic history.[5]
The surname (Déiseach in Irish) derives from the district of the Decies
(na Déisigh) in County Waterford on Ireland's south coast, which was
the clan's native territory. According to family tradition, the West Cork
branch sprang from a single refugee from that area, a pregnant woman
who, fleeing westwards from Cromwell's troops in their fearsome
advance through Leinster in 1649, eventually came to a halt at the sea at
East Barryroe, Co. Cork. There she gave birth to a son, the reputed head
of the entire West Cork Deasy dynasty.

The family prospered, acquired land, and by around 1700 had its
seat at Lisscrimeen Castle overlooking a secluded sandy beach and the
wide open sea beyond, the walls of which still stand. The Deasys inter-
married with notable local families, including that of The O'Donovan,
chieftain of an ancient clan who was a member of the Irish Parliament
of King James II. Later interesting family connections included Fitz-
James O'Brien, a well-known American journalist and an early writer
of science fiction,[6] whose mother (née Deasy) was an aunt of Mrs
Clerke; and possibly Edmund Burke, the great Irish politician and
orator who (according to Mrs Clerke) was distantly related by marriage.[7] Whatever the details, it is undoubtedly true that the Deasy kin
had intriguing ramifications, and threw up into the world some
unusual and dashing characters.

A direct descendant of the original patriarch was the colourful
Timothy Deasy of Timoleague, Co. Cork, who was arrested in 1745 for
singing a Jacobite song and narrowly escaped the death penalty. When
the case against him as a Jacobite sympathiser failed, he was indicted on

the arms charge, since Catholics were forbidden from possessing arms. Meanwhile, his elder brother had made his fortune in Jamaica, where he acquired valuable estates. The proceeds of the Jamaican property, a major source of the family wealth, were eventually inherited by Timothy's grandson Rickard, Agnes Clerke's grandfather. As a boy, Rickard attended a school in Cork – such schools, run by private teachers, were very common – where he clearly achieved a high standard of literacy and polish, as revealed in the memoirs he wrote for the benefit of his children in his old age.[8]

The Deasys were Roman Catholics and traditionally nationalist in politics. On the repeal of the Penal Laws Rickard Deasy achieved the distinction of becoming in 1793 the first Catholic magistrate in Cork for a century. In 1807 he set up a brewery in Clonakilty – Deasy and Company – which became the town's chief industry and operated until 1940 when it was succeeded by the present Deasy Bottling and Mineral Water Plant. The soaring chimneys of the brewery still dominate the skyline of Clonakilty: the old structure is in fact being conserved for its historical interest. The brewery incorporated its own cooperage with highly skilled craftsmen: a magnificent polished oak cask with brass hoops conveyed the firm's prizewinning stout to the International Exhibition, commemorating the 400th anniversary of the arrival of Columbus, in Chicago in 1896.[9]

The Deasys also established another highly successful business as sea-merchants and shipbuilders.[10] Deasy's Quay, still so named, in Clonakilty Harbour, shows the remains of the docks and of the shipyard which flourished there in the first half of the nineteenth century. Several schooners were built there, including the *Mary Ann* (the name of Deasy's wife and of one of his daughters) and the *Catherine*, named after the daughter who was to become Agnes Clerke's mother.

The commerce at Deasy's Quay included smuggling, chiefly of wine and brandy from France, a lucrative and far from disreputable trade in that part of the country. The O'Connells of Derrynane, to which Daniel O'Connell belonged, famously belonged to the same smuggling consortium which operated along the south coast.

Rickard Deasy and his wife Mary Ann, whom he married in 1802, were people of considerable influence in the community, which they exercised to the full in charitable and social causes.[11] The West Cork Regional Museum at Clonakilty devotes a special section to the history of the Deasy family, spanning four generations and ending with Henry Hugh Peter Deasy, Agnes Clerke's first cousin, a noted explorer in Tibet at the end of the nineteenth century.

The immediate maternal family

Rickard Deasy's elder son, who adopted the alternative spelling of Decie as his surname, settled in England. He died at the age of only 29, leaving a successful family with its own share of adventure: one of his sons was a noted transatlantic yachtsman.[12] The younger son, Rickard Morgan, the dominant member of the family in that generation, was a man of exceptional intellectual talents. He entered Trinity College Dublin at the age of only 15, graduated in law at 20 and was called to the Bar at the unusually early age of 23. At the time of the Clerkes' marriage Rickard Deasy was practising as a barrister in Dublin at the start of a highly successful career.

Rickard Deasy and John Clerke, who entered Trinity the same year at the even younger age of 14, were fellow-students for five years. It was through their friendship that Clerke was introduced to Deasy's favourite sister: according to their mother, Rickard was the 'idolised brother' of her 'dear affectionate child', Catherine.[13]

Catherine was the youngest of six daughters. The others were Anne and Margaret, who became nuns; Honoria and Mary Anne who, like Catherine, made marriages to husbands in Co. Cork; and Ellen who died young, probably while still in her teens.[14] The girls were educated at the Ursuline Convent, Cork, where their names may still be seen in the school roll.[15] They also had the benefit of a learned tutor, John Sheahan, employed by the family to teach the children, who went on to become a journalist with the local *Southern Reporter* and other papers.[16]

Convent boarding schools grew in popularity in Ireland throughout the nineteenth century. The Ursuline nuns, one of the first religious orders to set up establishments in the country after the abolition of the Penal Laws, were well-known educators of girls from middle class Catholic families. Though women of that era were not destined for careers, the Ursuline nuns imparted a high standard of education to their pupils. The subjects on the curriculum were English, French, Spanish, Italian, History, Geography, Religious Knowledge, Music and Art – the usual list of accomplishments of young ladies of their time – though Needlework is strangely missing. Science was on offer, consisting probably of a little Nature Study, but Mathematics was not mentioned.[17] Parents were promised that 'every gentle and persuasive method shall be used to impress on the minds of young ladies an elevated and habitual sense of decency and propriety, and to polish and refine their manners'.[18] Music was very special to Catherine. She practised the piano every day of her life right into old age. She also played the harp, and in her seventies still enjoyed performing Irish melodies and accompanying her daughters' singing.

The Ursuline Convent in Cork was one of the places visited by the novelist William Makepeace Thackeray in the course of his tour of Ireland in 1842.[19] Already prejudiced against the idea of an enclosed religious life, he produced a sarcastic account of conditions within the convent walls. He concluded that 'we have as much right to permit Sutteeism in India, as to allow women in the United Kingdom to take these wicked vows, or Catholic bishops to receive them'. Sister Josephine (formerly Anne Deasy, the eldest of the six sisters) who was a member of the community there since 1822 would have given a different interpretation.

The Clerke–Deasy marriage

John William Clerke and Catherine Deasy were married on 9 July 1839. She was the last of the daughters to marry. In Ireland's religiously divided society, such an alliance was sometimes seen as giving a Catholic partner access to the Protestant 'gentry', but this was far from the case with the Deasys, who were wealthy and independent. The Clerkes belonged to the category of 'liberal Protestant' frequently mentioned and praised by Catherine's father in his extant recollections of family and political life. A marriage settlement of some complexity worth £1,500 was drawn up on Catherine, 'written on sheets of parchment like mainsails'.[20] In the course of time John Clerke would be further helped by his more affluent brother-in-law Rickard.

According to the civil rules then prevailing, the Clerkes' marriage, being mixed, was officiated twice, once before the Church of Ireland (Anglican) clergyman, and afterwards in the Catholic Church in Clonakilty before the Reverend David Walsh, Parish Priest. The bride's father no doubt intended to memorialise the occasion when he named the schooner *Catherine*, a vessel of 87 tons built in his shipyard in 1840. The *Catherine* was lost off the Scilly Islands in 1850 when on her way to London with a cargo of oats. She filled and sank as her pumps were unable to cope.

The Clerkes' three children – Ellen Mary (born 26 September 1840), Agnes Mary (born 10 February 1842) and Aubrey St John (born in April 1843) – were all brought up in their mother's Catholic faith, and remained active and exemplary members of their church throughout their lives. In 1840, only one year after their marriage, John's father, the devoted Dr St John, died at the age of 72. At about the same time, certainly before 1845, Catherine's parents left Clonakilty and went to live abroad. It appears that – perhaps as a result of their lavish and over-generous life-style – they ran into financial difficulties and emigrated to the Island of Jersey. They resided in Jersey for the rest of their days while keeping actively in touch with home. Rickard Deasy died in 1852, and his wife in 1853.

Skibbereen

The small town of Skibbereen[21] on the beautiful remote south coast of Ireland where the Clerkes reared their family had long been a focus of business for the surrounding countryside which, according to *Pigot's Directory* of 1824, was 'thickly inhabited by an opulent gentry'. It had its weekly market where farmers brought their produce and regular fair days for the sale of livestock. A coach passed through each morning and evening, carrying mail and connecting travellers to the neighbouring towns and to the city of Cork, a distance of 53 miles. There was a court-house, a police barracks, a post office, a coastguard station, the offices of the famous *Skibbereen Eagle* newspaper, hotels, public houses, a medical dispensary and the inevitable workhouse for paupers, erected in 1842. The Bishop of Ross had his seat in Skibbereen, where a hand-some Catholic Cathedral had been built in 1826. Abbeystrewry Church of the Church of Ireland dated from 1827 and the Methodist Chapel from 1833. There were schools attached to these churches, the largest being the Catholic school supported by the National Board.

In the early years of the nineteenth century the town, of about 4,000 inhabitants, had been a thriving centre of the textile industry, producing woollen and linen goods in its mills. After the Napoleonic Wars that industry fell into decline, and with it the overall prosperity of the populace. There still existed a large brewery and a steam corn mill; craftsmen such as coopers, rope-makers and dyers followed their trades amid the crowded shops on the town's four streets. Much of the popula-tion, however, according to the reports of various travellers, was clut-tered in miserable dwellings in the surrounding lanes.

The bank, a relatively new feature in Irish provincial towns, was sure to be a focal point in local commercial life. The three-storey build-ing, which today houses business offices, had living quarters for the manager and a plot of land behind leading down to the river Ilen (Figure 1.2). The family and bank customers shared an entrance door which opened directly from the street into a large hallway. Off this on one side was the bank office with a window facing the street. The private part of the house was shielded by a pair of doors, behind which is still preserved the original curved mahogany staircase leading past a huge window commanding a beautiful view of the river. The same panorama of the river and the barge traffic could be enjoyed from the top floor.

The Great Famine

Then came the calamitous Great Famine of 1845–50. The potato blight which caused the famine first struck in Co. Cork and quickly spread. The area around Skibbereen, noted for the excellent quality of its pota-

Figure 1.2 Agnes Clerke's birthplace (with white doorway) in Skibbereen.

toes, the staple food of the majority of the local people, was one of the worst afflicted in the whole country;[22] by the end of the decade a third of its population was lost, either through death or emigration. The only 'growth industry' in the town was the conversion of disused buildings to form extra workhouses; in the ten years from 1841 over 4,000 people died in these institutions, not counting many who died in their homes, on the roadside and even on the streets of Skibbereen. In the countryside there were wholesale evictions of tenants unable to pay their rents, and a consequent exodus of those fit enough to depart for America. There were numerous instances of rioting and theft of food, followed by court proceedings and sometimes transportation.

Living as they did in the midst of the community, the Clerke family could not escape the realities of life about it, and, in fact, had a very direct experience of the worst phase of the Great Famine. At the first signs of impending tragedy, John Clerke, as his father and other members of his family had done in earlier crises, took measures to alleviate the desperate plight of the hungry. In the autumn of 1845 a group of townspeople set up a soup kitchen in the steam mill and opened a private relief fund with Clerke as treasurer and secretary. When government action belatedly came into play, Clerke, as secretary of the Committee of Gratuitous Relief, was the one who communicated with officials in Dublin, pleaded for government assistance, and took charge of public relief subscriptions.

Hunger was not the only evil stalking the country. Famine brought with it a cholera epidemic which was not confined to the malnourished but spread through the professional and business classes. John Clerke at the Bank was one of those who fell victim; so did the police inspector next door, the doctor's family, the hotel-keeper and other neighbours, some of whom died.[23] Clerke's superiors sent a replacement from London to manage the bank who on arrival in March 1847 sent a report back to London on the terrible conditions he encountered – dead bodies carried away on carts, people dying of fever on the streets, the bank office besieged by skeletal beggars. The shocked Directors in London passed the officer's report to the head of the British treasury 'for your information; but you are at perfect liberty to make such use of the communication as you shall consider to be proper'. This harrowing account was put on official record, to be quoted in many accounts of the Irish famine.

Mr Clerke took two months to recover from fever. It is unlikely that the children, being all under the age of six, would have retained a clear memory of this particular anxious episode. Yet they cannot have grown up unaware of the general situation around them which went on for some years, and of their father's continued charitable exertions. Pat Cleary,[24] a local historian, writes: 'How Mrs Clerke shielded her children from the horrors of the famine is difficult to fathom. They lived at that side of town where there were two soup kitchens, both visible from the Clerke residence. There were other workhouses in the town which they would have passed on their way to Mass'. At the less prosperous end of the street, deserted small dwellings formerly the homes of poor families, lay derelict, while as late as the end of the decade the work-house was still packed with 2,500 people.

In 1854, the Crimean War broke out. Though geographically happening far away, it was a significant occasion in the south of Ireland, where many soldiers had enlisted in the army. Agnes later put on record how a display of the northern lights seen in the autumn of 1854 was believed by the people of nearby Berehaven to signal the deaths of those slain at Balaclava.[25]

Along the street from the Clerkes was the grocery and hardware shop of Jeremiah O'Donovan Rossa, Skibbereen's most famous son who was founder of the Phoenix Society, a political and literary club of local young men with nationalist aspirations, which in 1858 became incorporated in the Fenian movement.[26] Rossa's arrest and the subsequent police activity would not have passed unremarked by the teen-aged Clerke children. They were too old, however, to coincide with another celebrated inhabitant of West Cork, the writer Edith Somerville (born 1858), who with her cousin Violet Martin formed the famous literary partnership of Somerville and Ross and were the authors of several novels and popular stories with a local flavour.

The family circle

Despite the shadow of the Famine, Aubrey Clerke in middle-age had the happiest recollections of his childhood.[27] He recalled a close devoted family where reading, study and music were the children's occupations, under the tutelage of their gifted parents. The father was a shy and scholarly person, more interested in intellectual pursuits than in 'mixing in society',[28] a temperament inherited quite particularly by Agnes. Mrs Clerke, a sympathetic and high-minded woman, was more outgoing by nature: Ellen, who was fond of riding and boating, had rather more of her mother's lively personality.

At the age of 12 Aubrey was sent to St Patrick's College, Carlow, a Catholic secondary school under the aegis of secular clergy. St Patrick's College, which still flourishes, had a senior seminary for students preparing for the priesthood and also instructed senior pupils for the

London University external examinations. Under normal circumstances, Agnes and Ellen might have been sent to board at the Ursuline Convent in Cork, where their mother had been educated and where their aunt was a senior member of the religious community. That this did not happen may be put down to Agnes being a delicate child and the sisters not wishing to be separated. The more robust Ellen remained Agnes' companion and protector all her life; indeed, Agnes was the darling of the family, to whom Aubrey, too, remained utterly devoted.

The parental decision to instruct their daughters all by themselves was a tremendous advantage for these exceptionally intelligent and diligent girls who quickly attained a phenomenally high level of education. The parents' effort was all the more remarkable when one considers that the father was not a man of leisure or of great wealth: he had his work at the bank and his self-imposed civic duties.

As they pursued their studies and grew to adulthood, the young girls were not without encouraging role models within the family circle. Their aunt Anne (Sister Josephine) in the Ursuline Convent in Cork, where she spent over 50 years, was, according to the convent records, a woman of exemplary character, 'endowed with splendid talents and possessed of an exceptional amount of information'.[29] Another aunt, Margaret (Mother Vincent), was a religious in the Convent of Mercy in Cork, having entered the order in its early days as a young novice in 1835. The Order of Mercy, founded in Ireland in 1831 and dedicated to the needs of the poor, expanded rapidly throughout the English-speaking world; by the end of the century hardly a town in Ireland, Britain and North America was without its Convent of Mercy. The fearless sisters also gave noble service as nurses in the Crimean War and in the American Civil War. To some extent Irish nuns in the nineteenth century were more emancipated than contemporary non-Catholic Englishwomen, with better opportunities to lead serious and active lives. A leading feminist, Barbara Bodichen, thought it 'happier by far a Sister of Charity or Mercy than a young lady at home without work or a lover'.[30]

Mother Vincent's mission was to England, where in 1843 she established a convent in Sunderland, an industrial town with its quota of poverty and misery. She and her companions worked in hospitals, tended the sick, opened schools and founded a convent which still operates. Mother Vincent returned to the Order's house in Cork, where she lived out a life of great saintliness and austerity until her death in 1878. In those last years she gave service through her writings, including translations into English of Spanish devotional works.

Another admirable female member of the family was Catherine Donovan – Mrs Clerke's first cousin and Ellen's godmother – educational pioneer and founder in 1819 of a school for local girls in Clonakilty. Compelled by poor health to abandon a vocation to the religious life, 'Miss Kitty', as she was affectionately known, resolved to devote her life to good works.[31] The Clonakilty School of Industry taught reading, writing, and useful crafts like needlework and knitting, and acquired a reputation well beyond the shores of Ireland for its beautiful embroidery and lace-work. Despite the setback of the Great Famine, the indomitable Miss Donovan with help from friends and members of the extended Deasy clan kept the school afloat up to her death in 1858. Hundreds of girls who passed through the Clonakilty School during the course of its existence were enabled to earn their own living at home, in Britain or in America.[32]

To Agnes and Ellen, as to these women, hard work came naturally, and a love of learning and intellectual achievement were not regarded as beyond the bounds of a woman's vocation.

2 Ireland and Italy

Growing up in Skibbereen

John Clerke, a man of profound all-round learning, was to his children a painstaking teacher, competent to instruct them in Latin, Greek, mathematics and the sciences.[1] On the practical side, he had a chemistry laboratory in the house where he performed experiments, and a telescope mounted in the garden through which the children were sometimes treated to views of Saturn's rings or Jupiter's satellites.

Astronomy for him was more than a hobby. The four-inch telescope, probably a portable transit instrument, was equipped with a chronograph for timing the transits of stars across the meridian. With this arrangement Clerke was able to provide a time service for the town of Skibbereen, which was as yet unconnected to the outer world by either railway or telegraph.

The principle of timekeeping by the stars is that the astronomer, by referring to a catalogue of star positions, knows the exact instants when these stars cross the southern meridian in the sky each day or night. The time thus recorded is sidereal time, which the astronomer, again by use of the almanac, is able to convert to local mean solar time, in this case Skibbereen time. This in turn could be converted to Dublin time (the standard Irish time, itself 25 minutes behind Greenwich mean time) by allowing for the difference in longitude between the two places. The effort required to carry out this procedure was not negligible, as transits of stars had to be observed frequently to maintain the clock's regularity. The clock itself would have been displayed within the bank or in a window where it could be consulted by the public. This valuable service to the community is another example of Clerke's civic spirit, and one which he took seriously: he is attributed with the profession of 'astronomer', among others, in the genealogy of the Deasy family. All trace is lost, unfortunately, of the telescope which played its part in Agnes' childhood.

Agnes as a young child became fascinated by astronomy. 'I had an immense love for the subject, and love begets knowledge', she said in later life.[2] Ellen preferred to write poetry. Verses of hers composed when she was only 13 – a romantic ballad in the style of Walter Scott – survives in a collection of her work published in 1881. But though they had their preferences, both girls were keenly receptive to every subject that came their way, and shared all each other's thoughts and interests. Latin and Greek were no chore; in adult life Agnes would read Homer for recreation and mental refreshment. Languages came easily to them: their mother probably introduced them to French and Spanish, and undoubtedly taught them music, their lifelong joy. Those feminine occupations of middle-class young ladies, sketching and needlework, were absent from the syllabus for her daughters. It is notable that neither of the Clerke sisters acquired an interest in art. Though both wrote widely on many subjects, art was never one of them.

The family library

Fortunately, it is possible to reconstruct fairly well Agnes' early education. Part of John Clerke's library, including several books on astronomy, survives in the Deasy family.[3] It can only be a remnant, however, since it contains very few Latin (and no Greek) works such as one would expect to find in the library of a classicist. The many books that Agnes and Ellen undoubtedly acquired in their adult lives are also lost, perhaps given away by their brother after their deaths.

The earliest books, dated 1831 (the year when at the age of 17 John Clerke became a scholar at Trinity College Dublin), are Nathaniel Hooke's Roman History (1767), Anecdotes of Distinguished Persons in four volumes (1831), Goldsmith's Works (1831) and a translation of Cervantes' Don Quixote. Other volumes include the works of Swift, Shakespeare and Milton. The library was, one may assume, originally stocked with the standard works of English literature.

Of the scientific books, the much annotated *Elementary Treatise on Astronomy* by Woodhouse (1812) dates back to Clerke's university days. In 1840 (the year after his marriage) Clerke acquired J.P. Nichol's two well-known books, *The Phenomena and Order of the Solar System* (1838) and *Views of the Architecture of the Heavens* (third edition 1839). These were followed by Alexander von Humboldt's *Cosmos* (Otté's translation) (1849), O.M. Mitchel's *Orbs of Heaven* (1851) and Dionysius Lardner's *Museum of Science and Art* (1855). Other books, now missing but mentioned by Agnes herself, were Joyce's *Scientific Dialogues* and John Herschel's *Outlines of Astronomy*.

According to Agnes' own recollection, Joyce's *Scientific Dialogues* was the first of these books to come her way. It was a popular dictionary of science which had been reprinted several times since it first appeared in 1807.[4] Nichol's excellent books would have also inspired her at an early stage. John Pringle Nichol was Professor of Astronomy in the University of Glasgow and a populariser in the best sense of the word, who combined 'rhetorical power with exact knowledge', to use Agnes Clerke's own description of his talent.[5] He was educated at the University of Aberdeen, where he achieved the highest honours in mathematics and physics. He made himself a successful career in schoolteaching and public lecturing before being appointed to the Regius Chair of Astronomy in Glasgow in 1836.

Nichol's professorial duties in Glasgow were not onerous. He found time to write several books which, to quote Agnes Clerke again, were 'eloquent, enthusiastic and learned'. His book on the Solar System was a straightforward account of the sun and planets, with no mathematical encumbrance, a very suitable point of departure. The *Architecture of the Heavens*, his first book, published in 1838 (with nine further editions before the 1850s) was his best loved. It was, perhaps, the earliest popular exposition of that subject of perennial appeal, the universe on a grand scale, what we now call cosmology. It discussed William Herschel's model of our stellar system and Laplace's theory of the origin and probable destiny of the 'material creation'. The book was free of technical terminology and illustrated with drawings of Herschel's nebulae.

Architecture of the Heavens was subtitled 'a series of letters to a lady' and 'respectfully inscribed' to a lady, Miss Ross of Rossie. A book addressed to women was in those days shorthand for a popular book, reminiscent of Mary Somerville's *Physical Sciences* published four years earlier, which aimed at making the laws of nature 'more familiar to my countrywomen'.[6] In the third edition of *Architecture of the Heavens* (1839), which is the one that Agnes Clerke read, Nichol had evidently forgotten the ladies. In his preface he noted that the huge sales of the first two editions had encouraged in him the hope that he had 'succeeded in familiarising many of my countrymen with the sublime truths unfolded by the higher Astronomy'. The novelist George Eliot (Mary Ann Evans) fell under the spell of Nichol's books. She wrote in 1841:

> I have been revelling in Nichol's *Architecture of the Heavens* and *Phenomena of the Solar System* and have in my imagination been winging my flight from system to system, from universe to universe, trying to conceive myself in such a position and with such a visual faculty as would enable me to enjoy what Young enumerates among the novelties of the 'Stranger' man when he bursts the shell to 'Behold an infinite of floating worlds/divide the crystal waves of ether pure/in endless voyage without port.[7]

Nichol's *Architecture of the Heavens* undoubtedly captured Agnes Clerke's imagination as well. It foreshadowed her *System of the Stars* of 1890.

Orbs of Heaven was the work of a respected American academic astronomer, Ormsby MacKnight Mitchel, professor of mathematics and physics at Cincinnati College, where he founded an observatory in 1842, one of the earliest in the United States. He was also the founder of the astronomical journal the *Sidereal Messenger* which played an important role in the diffusion of astronomy in America. *Orbs of Heaven*, in a flowery enthusiastic style and with some fanciful interpolations, outlined the development of astronomy from Adam in the Garden of Eden on the seventh day of Creation through the watchers at the Pyramids of Egypt and at the temples of India, to the most recent discoveries. With not a single formula or symbol, the book would have been easily readable by the young Agnes. It may well also have triggered her interest in the historical aspect of astronomy: her decision to write a history of astronomy was already made, according to her friend Margaret Huggins, at the age of 15.

By the age of eleven, so she herself attested,[8] Agnes had mastered John Herschel's *Outlines of Astronomy*, a book she was to describe as 'perhaps the most completely satisfactory exposition of a science ever penned'.[9] This famous treatise was first published in 1849 and remained the most popular serious text on the subject for over 50 years.

It dealt principally with 'classical' astronomy – the sun, moon, planets, comets, stars, and the laws governing their motions – treated in considerable detail and requiring of the reader a knowledge of geometry and trigonometry. A section on advanced mathematical topics was intended for university students.

For physics, there was a volume on general science from the numerous works of the Irish-born Dionysius Lardner.[10] Lardner, Professor of Physics and Astronomy at the newly-founded University of London, was a versatile writer and editor of the *Cabinet Cyclopaedia*, which contained in all an incredible 133 volumes and attracted many famous contributors including Sir John Herschel.

The most difficult (from a student's point of view) of the books in John Clerke's library was Woodhouse's *Elementary Treatise on Astronomy*, a text designed for the serious student of mathematics. Robert Woodhouse, Lucasian Professor of Mathematics and then Plumian Professor of Astronomy at Cambridge, was the first to bring the methods of continental mathematicians to the notice of his countrymen.[11] Mathematics at this great university had long been languishing. In the field of calculus Newton's outdated and cumbersome notation was still in use. Woodhouse explained and advocated the continental notation in his book *The Principles of Analytical Calculation*, published in 1803. The great renaissance of mathematics in Cambridge was brought about by Woodhouse's three famous disciples, John Herschel, Charles Babbage and George Peacock.

Woodhouse's book on calculus was followed by others on different branches of mathematics, including, in 1812, the book in Clerke's possession. This covered spherical, positional and planetary astronomy in considerable detail, with the inevitable trigonometric formulae. All in all, a book demanding hard work and concentration. The copious underlines in pencil on John Clerke's copy shows that the book had been diligently studied by him in his student days. Whether Agnes studied it in equal depth is impossible to say with certainty. Notes in the margins, with important points labelled, may well have been aids to Agnes' homework.

John Clerke's surviving library contains no books on pure mathematics. However, being able to cope with Herschel's *Astronomy* (and perhaps also Woodhouse's) meant that Agnes acquired a competent working knowledge of mathematics, including trigonometry and algebra.

Since John Clerke always put a date in his books, we know that this well-chosen little scientific library was in place by 1855, when Agnes was only 13.

One book only from this era belonged to Agnes personally. It is inscribed 'Agnes Clerke from her fond Papa, 17 August 1855'. One opens it to find that it is not, surprisingly, about astronomy. It is T.B. Macauley's *Critical and Historical Essays*, in 2 volumes (1851 reprint) – an extraordinarily advanced and cerebral challenge to a child of 13. Macauley's famous Essays from the *Edinburgh Review*, written over a period of 20 years, covered political, literary and historical subjects of great interest and importance in their day. Her father's dedicatory words suggest that the book was a treat, an exciting gift. Agnes herself was to become a regular contributor to the *Edinburgh Review* some 20 years later.

John Clerke clearly did not consider Macauley too difficult either in language or in substance for Agnes, nor did he share the notion, still widespread, that women were not as intellectually capable as men. On the contrary, he quite obviously wished to stretch her mind, and if, being a girl, she could not enter university (as he had done at the age of 14), he would give her the opportunity to achieve the same result at home. She was clever, she was keen, she was determined, and she had an excellent and enlightened tutor in her father.

In the course of time some further scientific books were acquired. In 1859, before he went up to university, Agnes' brother Aubrey – who planned to study mathematics and physics – bought a set of Lardner's works, which have also survived in the family library. They were the *Handbook of Natural Philosophy* in 2 volumes *(Optics* and *Hydrostatics Pneumatics and Heat)* and the two-volume *Handbook of Astronomy* (1856–58).[12]

To these were added, in 1861, a set of books presented to John Clerke when the family left Skibbereen. The choice displays his unabated serious interest in astronomy. Two were by Robert Woodhouse: the *Treatise on Astronomy* Volume 1 (1821), a much expanded edition, intended for university students, of the book he already possessed; and Volume II, *Physical Astronomy* (1818), an advanced text dealing with the works of the great continental mathematicians Euler, Clairaut, Lagrange and Laplace. A third book was *An Introduction to Practical Astronomy* (1859) by Elias Loomis, Professor of Astronomy and Physics at the University of New York, a standard student textbook covering basic spherical astronomy, positional instruments, determination of time and latitude, etc. The range of books on astronomy now available to Agnes (aged 19) was comprehensive and quite advanced.

John Clerke also took an interest in microscopy (i.e. biology), which Agnes did not share, and already had two books on the subject in his library. The last book in the gift set was a beautifully illustrated *Treatise on the use of the Microscope* by John Quekett, a professor at the Royal College of Surgeons in London.

The leather-bound books, still perfectly preserved, are gilt-lettered and decorated. The outside cover of each volume is inscribed: 'A Memorial of regard and esteem from the Inhabitants of Skibbereen and its neighbourhood. August 1861'. They were the farewell gift to John Clerke, presented at a public gathering when he left the bank for a new career in Dublin.

John Clerke was now 47 years old, and, apart from his student years, had lived all his life in his native town. The Clerkes did not lose touch with their old home. Among other surviving books is a copy of Eugene O'Curry's monumental book on the Ancient Irish (published 1873), a gift to the family 'in memory of their unvarying kindness' from the Parish Priest of Rosscarbery in 1885, many years after they had left. It is also known that John Clerke acquired some land in the Skibbereen area which was inherited by Aubrey who eventually sold it to his tenants in 1914.[13]

The move to Dublin

Agnes was 19 and Ellen 20 when the next phase in their lives began, as their father took up an entirely new profession – in the Law. The post to which Mr Clerke was appointed was that of Registrar at the court of his brother-in-law, Rickard Deasy, newly appointed Baron of the Exchequer. Deasy had risen spectacularly in his profession, first as a barrister with a large practise in Dublin and then, having taken silk in 1849, as a judge on the Munster circuit. In 1855 he stood for Parliament in a by-election and was elected a Liberal Member for his native county of Cork. He held his seat by large majorities in successive general elections in 1857 and 1859. In the case of Irish members of Parliament at Westminster the description Liberal applied to all those who were reformers, though their allegiances were not all identical.

Daniel O'Connell, representing the Catholic and Nationalist cause, had achieved Catholic Emancipation in 1829 (which applied to Catholics in the whole of Great Britain) as part of a campaign which included the Repeal of the Union of Great Britain and Ireland. The Deasys were naturally supporters of O'Connell in the cause of emancipation. After emancipation, however, and after the death of O'Connell in 1847, not all his Irish followers continued to press for Repeal. Deasy was one of those who believed that, emancipation being achieved, the Irish cause was best served within the Union. He supported the official Liberal Party in the House of Commons, and held successively the posts of Solicitor General and Attorney General for Ireland. In the latter office he was responsible for an Act of Parliament, popularly called 'Deasy's Act' (1861), which attempted to solve one of Ireland's most intractable problems, that of relations between landlords and their tenants. It codified the vast amount of legislation on the issue, and endeavoured to put matters on a contractual basis. Deasy's Act, despite its theoretical fairness, turned out in practice, however, to be no more than a dead letter.

Deasy resigned his Parliamentary seat on assuming his new office in 1861 and took no further part in politics. As Baron of the Exchequer he was one of the country's first Roman Catholic judges. The office with its now abolished archaic title was equivalent to a High Court Judge. The Court of the Exchequer was located at the Four Courts in Dublin, that magnificent building on the banks of the River Liffey which is still the seat of the Dublin Law Courts. In those days a judge was responsible for supplying his own registrar. It was thus that John Clerke had the opportunity of starting a new career, one more in keeping with his intellectual turn of mind, and of transplanting his family from a remote country town to the sophisticated Irish metropolis.

The Clerke family home was at 12 Herbert Place, a spacious terraced house on the fashionable south side. Herbert Place is one of Dublin's most beautiful and dignified Georgian streets which still stands unspoiled. The Grand Canal, which was then a busy route for commerce, runs parallel to the terrace on the opposite side of the road. From the windows of the houses the inhabitants could watch the barges on the canal, and could take the same pleasant walks along the canal bank as the citizens of Dublin do today. The canal and its environs have strong associations with the literary world. A novelist of a later era, Elizabeth Bowen, was born in Herbert Place, a few doors away from where the Clerkes once lived; the canal bank nearby was a favourite haunt of the twentieth-century writer Patrick Kavanagh, to whom a statue and memorial bench have been erected in recent times.

The canal and its barges would have reminded the Clerkes of the River Ilen at the foot of their garden in Skibbereen. In other respects they would have found their new surroundings in Dublin a great contrast. No more customers in and out of the bank premises downstairs; no more busy main street traffic; no more weekly street markets or monthly cattle fairs. Inside the house, however, the accommodation was not very different from the house in Skibbereen. The hallway had the same sweeping staircase, illuminated by a large window, the drawing room on the first floor and the bedrooms above. In a design typical of many of the Dublin Georgian houses, the entrance was reached by a flight of outside steps, under which was the entrance to the lower floor at street level.

Herbert Place was not far from Judge Deasy's residence at 27 Merrion Square, the best address in Dublin. He had recently married, and the two families were to remain very close, though the Deasy children were considerably younger than their cousins. Deasy also owned a country house, the beautiful mansion of Carysfort, Blackrock, on the south side of Dublin Bay, an indication that he was now a man of considerable wealth. Nevertheless, he was far from ostentatious. He was noted in his professional life for his learning and judicial abilities, for his keen love of justice and freedom from prejudice. As a person, he was admired for his 'courtly and kindly demeanour',[14] a description which might well be applied to all members of the family.

Life in Dublin

Little is recorded of Agnes and Ellen's social life in Dublin. Agnes studied piano with a well-known teacher, Miss Flynn, one of the two sisters immortalised by the writer James Joyce in his story 'The dead'.[15] 'The Misses Flynn, teachers of pianoforte and singing' (as they continued to advertise themselves after they were married) had their establishment at 16 Ellis Quay. One of them, Margaret (1832–81), Mrs John Murray, was James Joyce's grandmother. The Flynn sisters were then younger than depicted by Joyce as Miss Kate and Miss Julia, but the company assembled at the Christmas party in his moving story gives a small glimpse of the Clerke sisters' social milieu as young women in the Dublin of the 1860s. In real life, the annual Christmas party at the Misses Flynn was organised by their pupils. The musical world in Dublin of those days was a very lively one, with many foreign as well as native artists.[16] Ellen probably took singing lessons from the second Miss Flynn, and sang songs of her own composition accompanied by Agnes on the piano.[17] Her instrument was the guitar; in her London years she was a pupil of Madame Pratten, a German-born musician whose flautist husband was for a time a performer at Dublin's Theatre Royal.[18]

One may surmise that the Clerkes did not neglect their studies and took advantage of Dublin's excellent libraries: the National Library in Kildare Street and the Royal Irish Academy in Dawson Street were within easy distance of their home. Their uncle the judge was a member of the Royal Irish Academy since 1846, with an entree to the city's intellectual circles. The signatures on his election certificate were headed by George Petrie, then Vice-President, and Sir Samuel Ferguson, two of the society's leading lights.[19] It was the heyday of celtic studies and Irish archaeology, when manuscripts, ornaments and artefacts were being discovered in abundance and added to the Academy's museum of antiquities.[20] Petrie was the dominant figure in the field. Ferguson, a lawyer by profession as well as an archaeologist, and his wife kept open house at their home in North Georges Street, where their hospitality included children's Christmas parties and Shakespearean readings by Academy members.[21] Another giant figure in this Golden Age was Sir William Wilde (his more famous son Oscar was still a schoolboy), surgeon, antiquarian and folklorist, who resided at 1 Merrion Square, close to the Deasys. His poet wife, who wrote under the name Speranza, was famous for her salon where she entertained writers and artists.[22]

The Clerke sisters may not have been involved in these social activities, but there are indications that they took an interest in the intellectual side of the Dublin world. Agnes refers in one of her articles to an aurora recorded in the *Annals of the Four Masters* (a history of Ireland written in the seventeenth century, transcribed and translated into English in the mid-nineteenth century).[23] Ellen alludes to Irish mythology in a book of her poems where she shows some acquaintance with the Irish language.[24] She also chose topics of Irish literary interest in a few of her essays.[25] Agnes no doubt continued to study astronomy during those six years in Dublin, helped by her brother, now a student at Trinity College.

Aubrey at university

The Clerke family's move to Dublin coincided with Aubrey's first year at Trinity College Dublin, where he became a scholar in mathematics and natural sciences. He had a brilliant academic career, gaining a string of awards and prizes. He graduated with a BA in 1865 with first class honours and gold medals in both mathematics and experimental science, and was awarded a seven-year postgraduate studentship in science. There was a distinguished school of mathematics and natural and experimental philosophy (i.e. physics) at Trinity College at the time. The undergraduate courses in mathematics included astronomy, of which the professor was the renowned mathematician Sir William Rowan Hamilton, until his death in 1865, though it is unlikely that he did much teaching in his last years. The students also read Laplace's *Méchanique Céleste*, a fundamental work on theoretical astronomy describing the motions of the bodies in the solar system in detailed mathematical form. Aubrey's postgraduate work earned him the MacCullough Prize in 1867 for an essay on a prescribed topic in theoretical mechanics.

The general principles of the astronomy involved in Aubrey's courses would have been known to Agnes from her study of Herschel's book; further advancement in this field required higher mathematics, including the use of calculus. Her essays on Laplace and other mathematicians written 20 years later demonstrate that she did indeed become familiar with these mathematical skills, though never an expert, possibly by keeping in touch with her brother's academic progress in those Dublin days. It would accord with what is known of Aubrey's kindly nature that he should share his knowledge with his sister. Throughout his life he was always at hand to take an interest in her work and help with mathematical problems.

A number of Aubrey Clerke's contemporaries at Trinity College became involved in astronomy in one way or another. The most conspicuous of these was Robert Stawell Ball, one year his senior at university, who became successively professor of astronomy at Dublin and Plumian professor of astronomy at Cambridge. Others were George Minchin, William H.S. Monck and John Ellard Gore, who made their principal livelihoods in other professions before making substantial contributions to astronomy as amateurs. The same remarkable period at Trinity College saw among its undergraduates the fourth Earl of Rosse, son of the constructor of the famous reflectors at Birr Castle and himself an excellent astronomer, and Howard Grubb, the world's leading telescope maker. Agnes Clerke knew these men in London after she herself became famous, but it is not known whether she had met them previously in Dublin. She is likely, however, to have known Thomas Luby, a mathematician and tutor at Trinity College (a student contemporary of her father who probably taught Aubrey), whose biography by her in the *Dictionary of National Biography* is partly compiled from private sources.

Aubrey Clerke, with his clutch of university honours, might have aspired to a career in mathematics or astronomy. However, opportunities in the academic world were few. A further deterrent in his case was that as a Roman Catholic he was barred from obtaining a Fellowship at Trinity College, the essential route to academic preferment. His contemporary Robert Stawell Ball recalled in his autobiography how he was on the point of taking up the law when the opportunity presented itself to be tutor to Lord Rosse's children, by which piece of good fortune he began his astronomical career.[26] It is therefore not surprising that Aubrey Clerke decided to follow his uncle and father into the law, though (possibly on account of the Fenian unrest in Ireland at that time) he chose not to seek his future in Ireland. He studied for the English bar (it was not necessary to attend courses in person), was called in 1869 and took up practice as a barrister at the Chancery Court in London. He specialised in land conveyancing, and acted for many English Catholic families.

By the time Aubrey moved to London his sisters had gone to live in Italy. The parents ceased to reside in Herbert Place in 1868, though the father continued to work as court registrar until Deasy's elevation to the Court of Appeal in 1878.

Italy

In 1867 Agnes and Ellen began a 10-year residence in Italy. At first, with their mother, they made extended visits to Rome and Naples, and from 1870 the family lived for seven years in Florence. Though the house in Dublin was given up, John Clerke retained his post, presumably returning to Dublin when the Court was in session. Summers were spent in the spa resort of Bagna di Lucca near the Mediterranean Coast, a favourite retreat of the large English colony in Florence.

Ellen has left some charming accounts of Italy and Italian life at that period.[27] They include descriptions of traditional religious festivities in Florence, celebrations of Christmas in Rome and of Easter in Naples, and village life and wine-making in the Tuscan Appennines. In April 1872 the Clerkes witnessed the eruption of Mount Vesuvius, recalled by Ellen in a scene in her novel *Flowers of Fire*, published 30 years after the event.[28] Agnes Clerke's only record of the occasion was a laconic: 'We witnessed the fire-drama it displayed in 1872, and are not likely to forget it'.[29] As far as is known, neither sister recorded her personal experiences of the dramatic public events of the unification of Italy or of the First Vatican Council convened by Pope Pius IX in 1870.

The years in Italy were a time of intense and serious study on the part of the two young women. They frequented the Florence city libraries, which were open freely to all readers, and educated themselves to an astonishing degree. Their interests ranged widely – Italian history and literature, the history of science, contemporary affairs, European languages, the classics. While Ellen was especially drawn to early Italian poetry, Agnes made a special study of the philosophy and science of the Renaissance. In Florence, a city teeming with reminders of Galileo, her reading concentrated particularly on the life and work of that great scientist and of his contemporaries (the complete edition of the writings of Galileo by Eugenio Alberi was available to her, while researches by other scholars were in progress and appearing at the time). As far as is known, she carried out these profound researches entirely without help.

Agnes continued her musical studies at the Istituto Musicale in Florence under its eminent professor of piano, Giovanni Buonamici, a former pupil of von Bülow in Munich. She became an accomplished pianist who kept up her music throughout her life. One of the few reported anecdotes of her time in Italy concerns an occasion when she and her mother, listening enthralled outside the apartment of the composer Franz Liszt in Rome, were spotted by the great man himself and invited inside. They heard him play, and Agnes, shy though she was, was prevailed upon to play for the company. The event, told by Agnes' great friend Lady Huggins, may not have been quite as spontaneous as recounted; the women would have been able to obtain an introduction through Buonamici to the aging composer, who was well-known for his hospitality to young admirers.

Agnes and Ellen left few clues as to their circle of personal friends in Italy. Agnes had one important contact in the English literary colony in Florence. She was Mrs Janet Ross (née Duff-Gordon), daughter of the writer and traveller Lady Lucie Duff-Gordon[30] and herself a writer of note, who with her husband Henry Ross lived in a beautiful villa near Florence called Poggio Gherardo. Lucie Duff-Gordon had belonged to the London literary set that included the Carlyles, George Meredith and Lord Tennyson. Her fascinating letters were published in her lifetime, but her daughter Janet edited and re-published them as *Letters from the Cape* and *Letters from Egypt* in 1875, while the Clerkes lived in Florence. It was Mrs Ross who gave Agnes Clerke her first introduction to the London literary scene, by recommending her to her first cousin, Henry Reeve, editor of the prestigious quarterly *Edinburgh Review*. Agnes had already written and published some minor pieces – which have not been traced – but the introduction to Reeve was her great breakthrough.

London

The Clerke sisters, now in their middle thirties, and their parents did not return to Ireland after their sojourn in Italy but in late 1877 or early 1878, on the father's retirement, settled in London with their brother, who began practice as a barrister with chambers in Lincoln's Inns Field. The Clerke family's first London home was 121 Finborough Road, Kensington.[31]

In 1884 the Clerkes moved to 68 Redcliffe Square, Kensington, which remained the family home until Aubrey's death in 1923.

3 London, the literary scene

The *Edinburgh Review*

The move to London effectively marked the beginning of Agnes and Ellen Clerke's prolific literary careers. Agnes' first articles in the *Edinburgh Review* were written in Italy and published in 1877.

The *Edinburgh Review*, described by Agnes Clerke as 'an organ of high critical thought'[1] was a quarterly journal devoted to literary, political and occasionally scientific subjects of topical interest. It carried lengthy reviews of recently published books or papers but went far beyond their immediate subject matter to wider general discussions. The contributors were anonymous. As its title suggests, the *Edinburgh Review* originated in Edinburgh at the beginning of the century but had long been published in London. Henry Reeve (Figure 3.1), its editor from 1855 onwards, was a distinguished man of letters and an influential figure on the London literary scene. The *Edinburgh Review* took a Liberal stance in political matters, as opposed to the Tory philosophy of its contemporary and rival the *Quarterly Review*.

Evidently keen to enter the field of journalism but unsure as to where to begin, Agnes Clerke offered Reeve two different topics from her Italian repertoire – recent politics and history of science – hoping, no doubt, that one or other would succeed. Reeve not only accepted both articles but asked for more. In reply to her request for advice for the future he wrote to her in Florence on 19 April 1877:

> My dear Miss Clerke, It gives me very sincere pleasure to have introduced you to your first literary success. I hope it may be the prelude to many more. I can hardly venture to recommend to you the course in which you should steer your bark. On scientific subjects I am very ignorant, but there has been an article in the *Review* on Spectrum

Figure 3.1 Henry Reeve, Agnes Clerke's first patron.

Analysis, by Professor Roscoe, and another on the transit of Venus last year. You have the advantage of seeing before your eyes the intellectual renaissance of Italy, and it has already supplied you with two very good subjects'.[2]

For the rest of her life Agnes Clerke provided two contributions yearly, usually in the spring and autumn issues, for the *Edinburgh Review*. In Reeve, who had in his youth lived in Switzerland, Germany and France, and was on intimate terms with literary and scientific figures in those countries and at home, Agnes Clerke found a most congenial associate. She also became a personal friend of Reeve and his wife; she and her sister and brother were their guests from time to time at their country home in Hampshire, and after Reeve's death in 1895 it was she who wrote his notice, and that of his father, a physician, in the *Dictionary of National Biography*.

Agnes Clerke's first article for the *Edinburgh Review* entitled 'Brigandage in Sicily' (April 1877) discussed the history of the region and the social conditions which led to the rise of the Mafia, a subject which no doubt would have surprised her later scientific readers. The recently published works under review were of course in Italian. The second article, 'Copernicus in Italy', published in July of the same year, was a discussion of pre-Copernican ideas in Italy based on volumes published in connection with the 350th anniversary (1873) of Copernicus' *De revolutionibus*. As is well known, the Polish Copernicus was the astronomer who first promulgated the theory that the sun and not the earth is the centre of the solar system. Such notions were already in the air when Copernicus was a young student in Italy, and it was this background that Agnes Clerke investigated in this, her first published article on an astronomical subject. The books under review were in Italian: one by the Galileo scholar Domenico Berti and the other by Giovanni Schiaparelli, the well-known astronomer and director of the Brera Observatory at Milan.

Agnes Clerke's principal contributions to the *Edinburgh Review* during the first few years dealt with the science and philosophy of the Renaissance, where she drew on the immense store of knowledge accumulated through her studies in Florence. The topics included the writings of Giordano Bruno and the philosopher Tommasso Campanella, while two successive essays, in 1879–80, dealt respectively with the writings of Francis Bacon (a new edition of whose *Novum Organum* had just been published) and with Isaac Newton and his precursors, including Robert Hooke and Jeremiah Horrocks. Not all the subjects given to her for review can have been as congenial to her as these. An article entitled 'Harvey and Cesalpino', following the tercentenary in 1878 of the birth of William Harvey, discoverer of the circulation of the blood, discussed five works, two in French, two in English including the Harveian Oration for that year, and a sixteenth-century medical text in Latin by the Italian anatomist Andrea Cesalpino. The writing of this excellent article, as is seen from her list of references, involved a study of history of medicine (in German) and modern medical textbooks including T.H. Huxley's *Elementary Physiology*. But Agnes Clerke was nothing if not thorough. She was also loyal, and could be depended on to tackle topics other than science for Reeve, ranging from current affairs in Africa and the Far East to Scandinavian antiquities.

Agnes Clerke's usefulness as an *Edinburgh* reviewer was helped by her extraordinary gift for languages. The books under discussion might be in Latin, Greek, German, French, Italian or Spanish as well as English, in all of which she was fluent. An article in 1878, for example, entitled *Gypsies*, which was concerned with philology, involved eight books, five in German, two in French and one in English. Another, on the history of Albania, in 1881, included a publication in modern Greek. On one occasion, so Lady Huggins tells us, Agnes Clerke taught herself enough Portuguese in six weeks to be able to undertake a study of a batch of books in that language and afterwards to read with enjoyment the whole of the *Luciad* in the original.

Altogether, over a period of 30 years, Agnes Clerke contributed fifty-four major articles to the *Edinburgh Review*, of which about a quarter were on non-scientific subjects. Though originally unsigned, her authorship is known from a list published in Lady Huggins' *Appreciation*. They are also identified in the *Wellesley Index to Victorian Periodicals*.[3]

Ellen and the *Dublin Review*

Meantime Ellen, too, was establishing herself as a writer. An article from her pen on the poetry of the artist Michelangelo appeared in the *Dublin Review* in October 1878, followed a year later by a similar article on the poet Dante. These were the first of regular contributions to that journal maintained for the rest of her life. The *Dublin Review* was a Catholic publication, founded in 1836 under the inspiration of Daniel O'Connell and the scholarly Nicholas (later Cardinal) Wiseman of Westminster, and given the title *Dublin* to indicate its Catholic connections. In format and content it was very similar to the *Edinburgh* and the *Quarterly* reviews, containing serious dissertations pertaining to new publications but including, of course, many of direct relevance to the Catholic Church. For the first 40 years of its existence the *Dublin Review* was dominated by Dr

Charles Russell, President of Maynooth College (Ireland's leading seminary for the Catholic priesthood), who was its co-editor until 1863 and a major contributor to its pages.[4] Unlike the *Edinburgh Review*, however, the articles in the *Dublin Review* normally carried the author's name.

Ellen soon became a dependable contributor to the *Dublin Review*, just as Agnes was to the *Edinburgh*, supplying up to four articles a year as well as shorter book reviews. From 1885 onwards she also provided the journal's regular 'Notes on travel and exploration', a feature with news about the latest findings in geography, geology and anthropology.

Ellen's material for her geographical contributions came, in many instances, from the reports of the Catholic Missionaries of St Joseph's Society in Mill Hill, London. That Society, founded in 1869 by Father Herbert Vaughan, now Bishop of Salford and eventually Cardinal Archbishop of Westminster, educated priests for the foreign missions in Africa and Asia.[5] Its reports were first-hand accounts of conditions in these sometimes little-known regions of the world, which Ellen would combine with her own scientific knowledge and her wide reading to produce informative and highly readable essays. By far the greater part of her *Dublin Review* articles – fifty in all in the course of her lifetime – were concerned with physical geography and economic conditions in foreign lands; she made geography, in fact, her scientific speciality. Mary Creese, in her study of women scientific writers of the Victorian era, has remarked on Ellen Clerke's perceptive commentaries on topics as diverse as the development of gold and diamond mines in Rhodesia and the opening of canals and railways for international trade.[6]

Herbert Vaughan became combined proprietor and editor of the *Dublin Review* in 1884, making the link between his missionary activities and the journal even stronger. Ellen Clerke, loyal and intelligent, was very useful to him – she is mentioned by his biographer as having sometimes helped him to prepare unsigned articles[7] – as she was to Cardinal Henry Manning whom Vaughan was to succeed in 1892. So well indeed did Ellen serve the *Dublin Review* that on the occasion of its Diamond Jubilee in 1896 the editor paid tribute to 'the indefatigable pen of Miss E.M. Clerke whose industry as a *Dublin* reviewer almost rivals that of Dr Russell'. A contribution singled out for special praise was an article, left unsigned, which she took over from Cardinal Manning, entitled 'The Destiny of Khartoum', on the tragic death of General Gordon in 1885. The story of Gordon was preceded by an impressive review of the history of the Sahara from the tenth century, the colonisation of Africa generally, and the labours of missionaries in more recent times. After the article appeared, Gordon's devoutly religious sister 'wrote to the writer to thank her for it, as expressive of her own feelings in the portion where Gordon's desertion is described'.[8]

Within a year of joining the *Dublin Review*, Ellen published the first of her series of six wonderfully evocative articles, already mentioned, on aspects of Italian life and literature in the famous literary *Cornhill Magazine*. As a writer for the *Cornhill*, Ellen found herself in the illustrious company of authors such as Robert Louis Stevenson. The magazine's editor was Leslie Stephen, later editor of the *Dictionary of National Biography*. Ellen's articles came to an end in 1883 – as did those of Stevenson and others – when the magazine was restyled as a vehicle for light reading. Further articles on life in Italy appeared in other journals.

Another leading literary figure in Ellen's world was Stephen's friend Richard Garnett, librarian at the British Museum, who shared her interest in Italian poetry. Garnett, as will be seen, was to become an important contact for Agnes as well.

Ellen and Agnes' lives thus ran in parallel, each carrying out similar commitments and meeting deadlines for their journals. It was quite remarkable how quickly they became established in the literary world – incidentally earning substantial fees, though their motive was never mercenary. Their brother Aubrey also did some writing; in 1878, the family's first year together in London, all three Clerkes were in print simultaneously in three quarterlies, Aubrey's contribution being one to the *Quarterly Review* on a political question in which he showed himself a staunch Unionist and Anglophile. In his own field, Aubrey was author and co-author of specialist treatises on aspects of Land law.

Encyclopaedia Britannica

When Agnes Clerke arrived in London it so happened that the publishers Adam and Charles Black of Edinburgh were bringing out the ninth edition of *Encyclopaedia Britannica*. As Reeve was a copious contributor to the *Encyclopaedia* it is not surprising that his protegée should have been invited to contribute on matters connected with the history of science. The twenty-four volumes of this major edition came out between 1875 and 1889. The early volumes were already completed before Agnes Clerke began writing, and it was not until volume 10 (1879) at the letter G that the first of her contributions, a long and erudite dissertation on Galileo, appeared. The main article on astronomy in volume 2 (which included an outline of the history of astronomy) was the work of Richard Anthony Proctor, the well-known populariser of astronomy who was probably also the author of the short unsigned entries on Tycho Brahe and Copernicus. It would appear that Proctor ceased to contribute at this point, to be replaced by Agnes Clerke.

Agnes Clerke's biographies were of an entirely different order from the two just mentioned. That on Galileo occupied 12 columns (about 9,000 words) and was backed by an exhaustive bibliography. It could well be said that this scholarly and even-handed dissertation of the life and work of that controversial genius is her finest piece of work, and still reads well after a lapse of more than a century. Her other major scientific biographies in *Encyclopaedia Britannica* were of the geologist–explorer Alexander von Humboldt, the optician Huygens, the astronomers Kepler and Leverrier, the chemist Lavoisier and the mathematicians Lagrange and Laplace. More than one commentator was of the opinion that the essay on Laplace was the most outstanding of these. It certainly was unique in that his work was treated mathematically, one of the few instances of its kind in all Agnes Clerke's writings and an example of her brother's unseen hand in her handling of difficult mathematical material. It is most likely that these particular essays owed their origin to Aubrey's prize dissertation on theoretical mechanics as a graduate student at Trinity College Dublin. Agnes' subsequent work shows that her own mathematical skills were less advanced, and it is significant that in the revision of the essay on Laplace for a later edition of the *Encyclopaedia* the difficult formulae were dropped. These contributions to the *Encyclopaedia* – which bore her signature – brought Agnes Clerke to public notice for the first time. Reviewers hailed her as the new Mary Somerville, the reference being to the Scottish woman mathematician celebrated for her treatises on mathematics and the physical sciences half a century earlier. 'It is worthy of remark', commented the weekly *Athenaeum*,[9] 'that the lives of Lagrange and Laplace have been entrusted to a lady, A.M. Clerke, who seems desirous to emulate the acquirements of Mrs Somerville'.

This comparison with the earlier woman writer was often to be made. Mary Somerville, almost entirely self-taught, had made herself master of the works of the French mathematicians of her day, who were then far in advance of their English counterparts.[10] She produced her own brilliant version in English of Laplace's famous treatise, for which she was admired by academics and approved of by the great Baron Laplace himself. She followed this with books for the general reader – as Agnes Clerke was destined to do – which remained popular for decades. It is interesting to reflect that Mrs Somerville resided in Florence for many years, but had left there before the Clerkes' time. She spent her last years in Naples where she witnessed and described vividly in her memoirs the same destructive eruption of Vesuvius in 1872 of which the Clerkes were also spectators.

After this batch of biographies, reaching letter L, published in 1882, there are no further contributions by Agnes Clerke to *Encyclopaedia Britannica* until the very last volume (1888), where a long and learned article under the title 'Zodiac' reveals her in one of her favourite fields – the history of ancient cosmologies and civilisations. The gap between the earlier entries and this one signifies the gestation and writing of *A Popular History of Astronomy during the Nineteenth Century*, her most famous work.

4 *The History of Astronomy*

The state of astronomy in about 1880

In those early years in London, while carrying out her many literary commitments, Agnes Clerke soon realised how far astronomy had advanced since she had studied it in Ireland.

That decade had been a time of unprecedented progress. Until the mid-nineteenth century astronomy had been principally concerned with the positions and movements of the heavenly bodies. Astronomers recorded with ever greater precision the paths of the planets against the background of the distant 'fixed' stars in order to understand their motions in space and to predict their future locations. They also searched for slight shifts in the positions of the stars themselves in order to discover their distances and their motions through space. For this purpose they produced star charts and compiled massive catalogues, but worried little about the nature of the stars themselves.

The motions of the planets around the sun had been elucidated by Isaac Newton in the seventeenth century in his great law of gravitation. Since Newton's time, more refined mathematical applications of that law were capable of explaining intricate details of the movements of the various bodies in the solar system caused by their gravitational influences on each other. Academic astronomers, who worked on these problems, tended to be mathematicians. Other astronomers specialised in studying the appearances of the sun, moon and planets through the telescope. The stars themselves, however, being enormously distant tiny points of light, appeared to be beyond the grasp of earthly observers.

In 1837 the distances of some of the nearest stars were determined for the first time through meticulous observations of minute

39

changes in their positions during the course of a year. Such an effect, called parallax, arises as a result of the earth's annual motion around the sun on a circle of 150 million kilometres radius, causing the relatively nearby stars to shift to and fro against the background of the very distant ones, as seen by the earth-bound astronomer. The shifts were tiny, a fraction of a second of arc at most, showing that the distances of even the nearest stars are immense: their light (travelling at 30,000 kilometres per second) takes years to reach us.

Once the distances of even a limited number of stars were known, it became possible to begin to build up a three dimensional picture of the distribution of stars in space. All such pictures make the sun just one member of a huge group of stars of various kinds, with – in the early versions – the sun more or less at the centre. In fact, even before the distances of the stars had been measured, the brilliant William Herschel working in England at the end of the eighteenth century had tackled the problem of the distribution of stars in space and had concluded that they were arranged in a system shaped like a flat disk or solid wheel. This 'millstone' model explained the phenomenon of the Milky Way, the hazy belt of light encircling the sky where stars are considerably more numerous than elsewhere.

Another of Herschel's discoveries was of double stars, pairs of stars in orbit about each other like the planets around the sun, proving that Newton's law of gravitation prevails in distant space just as it does locally – a fundamentally important demonstration that the laws of science are universal.

Such was the state of astronomical knowledge in the middle of the nineteenth century, presented to English-speaking readers in treatises such as John (son of William) Herschel's much reprinted *Outlines of Astronomy*, which Agnes Clerke had studied under her father's tutelage in Skibbereen.

The history and development of astronomy from earliest times up to that point was recorded in a famous *History of Astronomy* by Robert Grant, professor of astronomy at the University of Glasgow, published in 1852.[1] A quarter of a century later, when Agnes Clerke renewed her studies, the picture of the universe had taken on a new dimension. 'During the interval', said Agnes Clerke, 'a so-called "new astronomy" has grown up by the side of the old'. The 'new astronomy', later to be called astrophysics, had sprung up from the application of spectroscopy, and the far-reaching discovery that the chemical composition of the sun and stars could be read in their light.

The spectrum – the spread of colours – of sunlight is seen in its simplest form in the rainbow, where the colours are separated out by drops of water in the atmosphere. When examined in a more refined way using a prism to split the light into its component colours or wavelengths, the spectrum is found not to be perfectly continuous; a number of dark gaps are revealed where light appears to be missing or reduced. A chart of the spectrum showing these gaps or lines in the otherwise continuous rainbow had been drawn up early in the century by a German optician Joseph Fraunhofer, but the origin of these so-called Fraunhofer lines was still a great mystery. Were they an inherent property of the light itself, or were they caused by some absorbing matter in the path between the sun and earth, and if so, how did they come about?

It was already understood that light is transmitted in the form of waves, and that the colours seen by the eye represent waves of particular lengths. Spectra of hot gases investigated in the laboratory showed a great variety of wavelength combinations. Unlike the sun's spectrum, however, these consisted of bright lines, their wavelengths depending on the particular gases under examination. Each chemical element was known to emit its own individual and unmistakable set of colours or wavelengths – this was the foundation of the science of spectroscopic chemical analysis.

The breakthrough in the understanding of the sun's spectrum came in 1859 when the German chemist Gustav Kirchhoff and his colleague Robert Bunsen, a physicist, reproduced the spectrum of sodium gas in the laboratory in the form of dark as well as of bright lines. A sodium-burning flame on its own showed a bright yellow line (in fact a pair of lines). The same flame viewed against the light of a hot lamp showed dark lines replacing the bright ones. The glowing sodium vapour was in fact removing its own characteristic colour from the more intense light behind, producing artificially the phenomenon of the solar spectrum. Here at last was the explanation of the dark gaps in the sun's spectrum: they represented absorption of the light of the sun's hot surface by cooler gases in front of it, sodium being one of those absorbing gases. Further examination showed more coincidences. Other elements, starting with iron, magnesium and calcium, were soon recognised in the solar spectrum and interpreted correctly as belonging to the sun's own atmosphere. In the course of time almost all the known elements on earth were in this way found to be present in the sun. Beginning in the 1860s astronomers applied the same technique of spectrum analysis to the stars, with mounting success. The pioneers in this field were Giovan Battista Donati in Florence (whose work Agnes Clerke may have come across while she lived there, though there is no evidence that she ever met him personally), Father Angelo Secchi, a Jesuit astronomer in Rome, and William Huggins in London, who was to become Agnes Clerke's venerated friend. Secchi and Huggins became and remain the acknowledged fathers of astrophysics.

The discovery of identifiable spectrum lines in the stars was truly momentous. What had previously been believed impossible – that the chemical composition of the unreachable celestial bodies could be known to observers on earth – was now in the realm of feasibility. Astronomers could ask the question, 'what are the stars made of?' and expect to find an answer. 'That a science of stellar chemistry should not only have become possible, but should already have made material advances', wrote Agnes Clerke in the preface to her first book, 'is surely one of the most amazing features in the swift progress of knowledge our age has witnessed'.

The first book

The 1880 Autumn number of the *Edinburgh Review* carried an extensive review article by Agnes Clerke entitled 'The chemistry of the stars'.[2] The important material discussed was a work on *Spectrum Analysis* by Henry E. Roscoe, a leading laboratory spectroscopist (1869); Huggins and Millar's original paper on stellar spectra (1864) and Huggins' further one on the spectra of stars and nebulae (1868), a paper on the solar spectrum by Joseph Norman Lockyer (1879), Secchi's book *Le Stelle, Saggio di Astronomia Siderale* (1878), and R.A. Proctor's *The*

Universe of Stars (1878). Here Agnes Clerke traced the history of astronomical spectroscopy, a subject that must have been new to her when she began. Proctor's book, in which the case for a one-system universe was argued, also greatly impressed her. In fact, from this time onwards, she was never to change her view on the matter.

For the moment, however, spectroscopy, not the universe at large, was the main element in the 'new astronomy'. It was a discovery about which she felt there was a need, indeed a duty, to inform the public. 'The service to astronomy itself would not be inconsiderable of enlisting wider sympathies on its behalf', she wrote, 'while to help one single mind towards a fuller understanding of the manifold works which have, in all ages, irresistibly spoken to man of the glory of God, might well be an object of no ignoble ambition.' In this noble spirit she embarked on her first book, *A Popular History of Astronomy during the Nineteenth Century.*

Agnes Clerke began writing her book in 1881. It was 'arduous work – arduous, that is, to me, thrown as I am on my limited resources'[3], as she wrote to the American astronomer Edward Holden who was one of the very few outside her family who knew about her enterprise. Ellen, her devoted sister and lifelong champion who shared her every thought, commemorated the commencement of the great work with a poem describing its main topics in romantic terms. The poem was included in a collection by Ellen published that same year, which also contained the ballad 'Erline' written when she was a thirteen-year-old girl in Skibbereen.[4]

Agnes Clerke spent four years, excluding preliminary research, on *A Popular History of Astronomy during the Nineteenth Century.*[5] It was intended to update and complement Grant's *History* of 1852. One aspect of the 'new astronomy', wrote Agnes Clerke, had been 'to render the science of the heavenly bodies more popular . . . than formerly. It has thus become practicable to describe in simple language the most essential parts of scientific discovery.' The 'new astronomy' was indeed at that time almost entirely descriptive. By covering the entire century she could put it in context and set out in logical sequence the advances in our knowledge of the universe, from William Herschel's pioneering researches at the end of the eighteenth century to the time of writing.

The usual topics of numerous popular books – accounts of the solar system and the planets – were taken for granted; her account began where these left off. The description 'popular' in the book's title meant a non-mathematical treatment 'to enable the ordinary reader to follow with intelligent interest the course of modern astronomy' – though the result was far from light reading.

The 500 page *History* was remarkable for its extraordinary thoroughness. The author's breadth of knowledge, her capacity for assembling and collating data, were enormous. Being fluent in many European languages she could truly say that 'materials have been derived with very few exceptions from the original authorities'. 'The system adopted', she wrote, 'has been to take as little as possible at second-hand. Much pains have been taken to trace the origin of ideas, often obscurely enunciated long before they came to resound through the scientific world, and to give each individual discoverer, strictly and impartially his due.' The book was written in an attractive style, enlivened by engaging biographical sketches of the principal players in the saga. 'There are many reasons' she said 'for preferring a history to a formal treatise on astronomy. In a treatise, *what* we know is set forth. A history tells us, in addition, *how* we came to know it. The story to be told leaves the marvels of imagination far behind and requires no embellishment from literary art or highflown phrases. Its best ornament is unvarnished truthfulness, and this indeed at least may confidently be claimed to be bestowed upon it.' Like Grant's book, there were no illustrations or diagrams.

The preface to the book thanked just two people, neither of them local, 'for many valuable communications'. One was the American astronomer Edward Singleton Holden whom she had never met. The other was Ralph Copeland of Dun Echt private observatory in Scotland, formerly employed in Ireland, with whom she may have had contact earlier. However, correspondence with both men began only in the year before the book came out. The book was therefore to all intents and purposes entirely her own work, constructed according to her own plan, and fully living up to her ideal of a popular book for the intelligent reader. Apart from the correspondence with Holden and Copeland, her research was done entirely in the library of the British Museum.

The *History* published

The *History*, published by Adam and Charles Black of Edinburgh (publishers of *Encyclopaedia Britannica*) in December 1885, excited considerable comment. Readers wondered who Miss Clerke might be and again drew the comparison with Mary Somerville. The book, said one reviewer, 'is written not by a Fellow of the Royal Society, but by a gifted member of a scarcely less interesting association – for the narrative is traced by the pen of a lady on whom the mantle of Mrs Somerville seems to have descended.'[6] The *Dublin Review* made the same point, adding that the author was 'a young Catholic lady' (she was in fact 43).[7] Sir Robert Ball, who reviewed the book at length in the scientific weekly *Nature*, also remarked that the 'learned volume is a product of a lady's pen'.[8] Ball, already mentioned as a contemporary of Aubrey Clerke's at Trinity College Dublin, was now Professor of Astronomy at his old university and director of the College's Dunsink Observatory. His own *The Story of the Heavens*,[9] a highly successful popular book for mass readership, was about to be published. Agnes Clerke's book was of a different style. 'Few men of science who use this book', Ball wrote, 'will think that it ought to be classed as a popular work in the ordinary acceptation. It might be more correctly described as a masterly exposition of the results of modern astronomy in those departments now usually characterised as physical.' 'Miss Clerke's most admirable work fills a widely felt want. [Every astronomer] can in this volume obtain a vivid and accurate summary of what has been done, or, if he prefers, to read the original memoirs, he will be directed where to find them.' The book, in short, was astronomy for astronomers.

Edward Walter Maunder, an experienced member of the staff of the Royal Observatory, Greenwich, and head of its solar department, was of the same mind. 'There is always an especial pleasure', he wrote in the *Observatory* magazine, 'in seeing a strong and skilful worker engaged on a worthy and congenial task, and that pleasant spectacle is abundantly afforded by the book before us . . . The book will be full of interest to the general reader, for the story of the marvellous discoveries made in astronomy during the past hundred years is told in a felicitous and attractive manner; but it will not be less highly valued by the student and the astronomer, on account of its completeness and accuracy, and the really remarkable skill with which the leading points on which our knowledge has been increased . . . are seized upon and set forth.'[10] He concluded that 'it well deserves a place beside Professor Grant's *History of Physical Astronomy*, to which it will form an admirable supplement'.

An extraordinary aspect of the first edition of Agnes Clerke's *History* was that its author, living in London, should have worked for so long on such a weighty project unknown to the local astronomical community. Margaret Huggins, wife and collaborator of William Huggins, had long wondered (like the admirers of Sir Walter Scott in his anonymous days) who the 'Unknown' author of the article on the *Chemistry of the Stars* in the *Edinburgh Review* in 1880 might be. She was no longer in doubt. She instantly recognised that the *History* had come from the same pen, and said as much to Sir Robert Ball.[11] Both agreed, however, that the author was probably not a practical observer – a point made also by Maunder.

The History of Astronomy during the Nineteenth Century was an instant success and had to be reprinted within two months; an American edition was also brought out by Macmillans in the same year.

Amid the general acclaim, one ambiguous notice of the book must have mystified Agnes Clerke. It appeared in the popular illustrated scientific magazine *Knowledge*,[12] and was very short: '. . . though guilty of the questionable taste of conspicuously puffing a notorious self-advertising "astronomer", the authoress has, on the whole, given us a fair and pleasantly written record of the progress of astronomy during the last eighty or ninety years.' The riddle would be solved later, when she came to know of the feud between the magazine's editor, R. A. Proctor, and her friendly correspondent Edward Holden.

Meantime, among Agnes Clerke's *Edinburgh Review* articles were two of current scientific interest. One, 'Volcanoes and volcanic action' (April 1883) was an opportunity to describe the Vesuvius eruption of 23 April 1872 (which she had herself seen) as recorded by an expert eyewitness, Luigi Palmeiri of Naples Academy, whose account had been translated into English by the Irish geologist Robert Mallet in 1873. A range of other publications on allied topics, some historical, included Humboldt's *Cosmos* and Archibald Geike's recent famous *Textbook of Geology* (1882). This article was fortuitously topical, appearing in print just a month before the great Krakatoa eruption of September 1883. The second article, on 'Mountain observatories' (October 1884), was an up-to-date survey of experiences of observing conditions at high altitudes and included a description of large telescopes then in operation or under construction. Papers discussed included a general account by Edward Pickering of Harvard, Samuel Langley's experiments on Mount Whitney, California, and on Mount Etna, Charles Piazzi Smyth's earlier expedition to the Peak of Tenerife, Ralph Copeland's astronomical tests in the Andes, and the emerging new Lick Observatory described by Edward Holden.

5 A circle of astronomers

Edward Holden

It was Agnes Clerke's good fortune to have made contact, early in her career, with Edward Holden (Figure 5.1), whose help she acknowledged in the Preface to her *History*. Holden was at that time professor of astronomy at the University of Wisconsin and director of that university's Washburn Observatory. Agnes Clerke and Holden never met; they appear to have been put in touch by a mutual friend, Dr Richard Garnett of the British Museum, whom Holden had got to know on an earlier visit to England. The friendly correspondence between the two, initiated by Holden in 1884, was to give her a good start as a chronicler of astronomy, though Holden's own eventual career was destined to be less happy.

Edward Holden's visit to London took place in 1876 – a year before the Clerkes came to live there. Holden was at that time an assistant at the Naval Observatory in Washington – his first appointment – and had been sent by the US Government to inspect the Collection of Scientific Instruments in the Museum.[1] It would have been natural for him to visit the British Museum, and to make the acquaintance of Garnett there. Indeed, the two would have had much in common. Holden was a good linguist and a man of wide cultural interests among which was the deciphering of hieroglyphic stone writings in the Yukatan. He was later to publish papers on Persian poetry, and other oriental subjects.[2]

Garnett, one of a famous literary family, was Superintendent of the Reading Room in the British Museum, later becoming the Keeper of Printed Books and the editor of the printed catalogue – 'the best-known figure at the British Museum Library in his or any generation'.[3] He had

Figure 5.1 Edward S. Holden. Courtesy of Mary Lea Shane Archives of the Lick Observatory, University of California, Santa Cruz.

been educated entirely at home and at the age of 16 began work at the Museum library, where his father had earlier been Keeper of Printed Books. He remained there all his life. His store of information, his knowledge of books and his prodigious memory were legendary, as were his kindness and obligingness to readers and enquirers. An eccentricity of his was an interest in astrology in which he appeared to believe; he wrote a book on the subject, *The Soul and the Stars*, under the pseudonym A.G. Trent (an anagram of Garnett) in 1894, which is quoted as a reference in the article on astrology in the eleventh edition of the *Encyclopaedia Britannica*. Astronomy and astrology may well have been topics of mutual interest to Holden and Garnett when they met in London.

The Clerke sisters, as users of the British Library, would have naturally got to know Richard Garnett. A special link between Ellen and him was their shared enthusiasm for sixteenth-century Italian poetry.

On hearing, through Garnett, of Agnes Clerke's interest in astronomy, and that she was writing a book, Holden took upon himself the task of adviser. He wrote offering his help, and announced that a parcel of publications was on its way. Agnes Clerke was deeply touched by his generosity. 'It is an additional incentive to me to do my utmost to make my forthcoming volume worthy of the subject and [of] the aid from so eminent a quarter',[4] she wrote, ending her letter with greetings from Mr Garnett. The parcel arrived a few days later, containing his articles on the Trifid nebula (1877) and on the moons of Uranus, as well as his account of the total eclipse expedition he made to the Caroline Island in the Pacific Ocean in 1883. He also sent his report on the site survey for the proposed new mountain observatory in California – the future Lick Observatory – which was most timely, as Agnes was at that time writing her article on 'Mountain astronomy' for the *Edinburgh Review*.

> I have been writing a popular account . . . of the movement initiated in the United States for establishing mountain observatories, so that the Report of the Lick Trustees comes in most appropriately. Indeed, I do not know how sufficiently to acknowledge my obligations to you'.[5]

Holden continued to keep her supplied with information about his own work and with copies of his many articles. She was already familiar with his book on the Herschels, and his lengthy monograph on the Orion Nebula (1882). Holden had – to quote his biographer Donald Osterbrock – 'tremendous powers of reading, assimilating, and organising information',[6] a gift shared by Agnes Clerke herself. His writings, and his enthusiastic letters, were exactly what she needed.

Holden was of course one of the first to receive a copy of the *History of Astronomy* from its author. He was now a person of considerable influence, having just been appointed President of the University of California and director-designate of the new Lick Observatory on Mount Hamilton in California. (He was to give up the university appointment and become full-time director when the Observatory was completed in 1888.)

The Lick Observatory owed its existence to a wealthy patron, James Lick, who had set up a trust for the purpose of establishing a new modern observatory in California. On its high mountain site, the first ever to be purpose-built at a suitable location, and about to be equipped with an excellent 36-inch refractor, it promised to be one of the world's greatest astrophysical observatories.

'It is a truly enviable position for an astronomer, perhaps the most so in the whole world', wrote Agnes Clerke to Holden as soon as she saw the announcement of his appointment,[7] 'and I am especially glad that it should be occupied by you who have shown me such marked kindness. . . . Who can tell what wonderful experiences and discoveries are in store for you, when you get the 36-inch on the top of Mount Hamilton under the transparent skies of California? A thrill of hopeful expectation can scarcely be repressed at the thought.' Holden in his reply sent Agnes Clerke a copy of an anonymous review of her book in the *New York Nation* which she found 'most flattering' and 'evidently from the pen of a competent critic. Recognition in your great country is especially dear to English writers, and enables them to realise more forcibly the vastness of the audience they address. Should my *History* reach a second edition I hope to take advantage of the hints of my American critic, to whom I feel quite as much obliged for pointing out its faults as for emphasising its merits'[8]. It transpired that the reviewer was Holden himself, as he later confessed to her, to which she replied: 'I had not looked so high as to the Director of the Lick Observatory for the author of the notice of my book in the Nation . . . I had no idea who wrote it.'[9]

The close link with Lick Observatory from the first moment of its existence had a tremendous influence on Agnes Clerke, directing her gaze outwards towards large-scale astrophysics. When in 1889 Holden founded the Astronomical Society of the Pacific, a society open to professionals and amateurs alike, he placed a copy of her *History* in the library with a warm recommendation to members to read it[10] and enrolled her a member, its first from outside America.[11] The society's activities were recorded in the *Publications of the Astronomical Society of the Pacific*, a journal edited and largely controlled by Holden himself. Quite apart from providing the latest astronomical tidings, usually in advance of their publication, Agnes' pen-friendship with Holden took on a warm personal character. Her sister Ellen also joined in the correspondence. Their association was maintained unclouded until Holden left Lick in 1897.

Agnes Clerke learned about, but took little notice of, the spiteful aspersions being cast at this time on Holden – and on American astronomers generally – by Richard Proctor.[12] As already recounted, Proctor had written a one-sentence review of her *History*, which sneered at the 'puff' in her preface for a so-called 'astronomer' – by which he meant Holden. Proctor, a man of considerable intellectual talent, with a Cambridge training in mathematics, earned his living as a popular lecturer and writer, astronomy being his chief interest. The row began through Proctor's criticism of the planned Lick telescope – and of what he regarded as the craze for large telescopes everywhere – which he published in an American magazine and also in *Knowledge* in July 1887.[13] He cited various large telescopes in Britain that had not lived up to expectations. A 'fine example' of this was that of 'an esteemed friend' (obviously William Huggins) whose work with telescopes of 15 and 18 inches provided by the Royal Society had been greatly outweighed by his early work in 1864 with a small one. He then went on to name the failures of the 36-inch telescope at Lick. He forecast disappointment with the large telescope, and wished that someone 'could take away from those most able observers their big playthings and send them back to the smaller instruments with which they did such noble work'. Holden and others responded in kind; Proctor published his answers in the next issue of his own magazine *Knowledge*, where he declared that Holden was jealous that he, Proctor, and not Holden, had been selected to contribute to a certain encyclopaedia back in 1871.[14] Agnes Clerke dismissed the affair in a few words to Holden, written in May 1888: 'I sincerely hope you will receive no further annoyance from Mr Proctor. The Astronomical Society [with which he also quarrelled] rebuke to him ought to be final'.

Ralph Copeland

The only other person acknowledged by Agnes Clerke as having helped with her *History* was Ralph Copeland. An Englishman trained at the University of Göttingen, Germany, Copeland had been an assistant at the Earl of Rosse's Observatory at Birr between 1871 and 1874 and then, briefly, assistant at Dunsink Observatory, Dublin, before being recruited by Lord Lindsay (later Earl of Crawford) to take charge of his private observatory at Dun Echt, Aberdeenshire.[15] He was to become Astronomer Royal for Scotland and Regius Professor of Astronomy at the University of Edinburgh in 1888. From Dun Echt, Copeland visited South America in 1882 and explored the potential of the high Andes as a site for astronomical observing stations. Agnes Clerke's article on mountain astronomy for the *Edinburgh Review* in 1884 drew, among others, upon Copeland's recently published experiences. She corresponded with Copeland at that time, enquiring about the distribution of certain types of star (emission-line stars) in the southern Milky Way, as observed by him in the Andes.[16] Copeland was also known to Holden, whom he met on the latter's visit to London in 1876.

J. Norman Lockyer

As to direct contacts, the first astronomer whom Agnes Clerke encountered in person since coming to London was Joseph Norman (later Sir Norman) Lockyer (Figure 5.2). She made his acquaintance through the librarian at the Science Library. 'Hearing that there was a frequent visitor to the library who read nothing but books and papers on astronomy' Lockyer offered to help her with her studies.[17] The librarian can have been none other than Richard Garnett, who had also put her in touch with Holden. Lockyer was, of course, well known to Agnes Clerke by repute. His publication on the solar spectrum was one of those studied in her article on contemporary astronomy for the *Edinburgh Review*.

Lockyer, like Agnes Clerke herself, did not have a formal scientific education and had begun his career as a popular writer. He came to fame in 1868 when he devised a method of viewing solar prominences outside a total eclipse of the sun by the use of a spectroscope. His other notable success was his spectroscopic discovery in the sun of the chemical element helium, then unknown on earth. Lockyer was attached to the Royal College of Science in South Kensington, where he had an observatory and a spectroscopic laboratory for solar and laboratory

Figure 5.2 J. Norman Lockyer. Royal Astronomical Society.

work. In addition, he built in 1888 a private observatory attached to his house on the south coast of England at Westgate-on-Sea, where he used to organise parties for astronomical evenings. 'Among the many visitors to the observatory at this time', wrote Lady Lockyer in her biography of her husband,[18] 'none saw the wonders of the heavens with a more lively and intelligent interest than Miss Agnes M. Clerke, now well known for her excellent books on the history and problems of astronomy, but at that time a beginner, with only a few isolated articles to her credit'. The demonstrator at the telescope who gave her 'her first view of the sky through a large telescope – a privilege which she was able to appreciate to the fullest extent' was the young Alfred (later Sir Alfred) Fowler, one of Lockyer's assistants who went on to become the country's most distinguished astrophysicist.

Agnes Clerke made other interesting friends at Westgate-on-Sea. One was Andrew Ainslie Common, a highly successful amateur astronomer and photographer, maker of the famous 3-foot silver-on-glass reflecting telescope (later known as the Crossley reflector from the name of its subsequent owner) acquired by the Lick Observatory in California in 1893. A splendid photograph of the Orion nebula taken with this telescope in 1883 had earned for Common the gold medal of the Royal Astronomical Society. Agnes Clerke used a print of this famous photograph as a frontispiece to the second edition of her book, published in 1887.

Lockyer did not abandon journalism when he took up astronomical research. In 1869 he co-founded and became editor of the weekly scientific journal *Nature*, today the most widely distributed scientific journal in the world. When the question of publishing the *History* in America came up Lockyer warned Agnes Clerke of the danger of piracy in the USA, and recommended that she should find 'a friendly American astronomer' as a nominal co-author. She proposed the idea to Holden, though 'reluctant to encroach on [his] kindness', asking if he would provide 'a few sentences or remarks' for her second edition, and allow her to use his name on the title page in some such form as 'with additions by Professor E.S. Holden'.[19] Holden, however, did not respond to this. The American edition was brought out in 1886 by Macmillans, the publishers of *Nature*.

Lockyer now encouraged Agnes Clerke to write another book, and even suggested a subject for it. 'The more I think over the book on the Development of Spectrum Analysis, the plan of which you were so good as to sketch out for me, the more ambitious I feel of some day writing it', she replied (in July 1886), 'but the more strongly I also feel that I must wait and learn much before I can hope to be competent to the undertaking'.[20] In the meantime, she was considering a book on 'Stars and Nebulae' and intended to write to Macmillans asking for their opinion from a commercial point of view.[21] Whatever may have been their reply, this idea was not followed up; instead she went on to write *The System of the Stars* (1890). She agreed to read the proofs of Lockyer's own forthcoming books, *Chemistry of the Sun* and *Movements of the Earth* (published 1887) – 'very proud of the confidence you show in me' she wrote, promising to 'study the sheets most carefully'.[22]

A second edition of the *History* was soon called for. When this was ready in 1887 Lockyer suggested that Agnes Clerke write her own announcement of it in *Nature*. 'It is very kind of you to allow me the space', she said, 'and of a piece with all your friendliness to my book and myself.'[23] The note read: 'We have received the second edition of Miss A.M. Clerke's *History of Astronomy during the Nineteenth Century*. We regard it as a most encouraging sign that in a period of not over 18 months the first edition should be exhausted. No book is likely to foster the love of the subject among people who are interested than Miss Clerke's'.[24] Was Agnes less modest than she outwardly appeared, or did Lockyer append the last sentence?

Lockyer thus played a significant part in Agnes Clerke's early success and in drawing attention to her talents. She became a copious contributor to *Nature*, writing expert reports on new developments in astronomy and book reviews, including those of foreign language publications such as a volume of Kepler's correspondence in German, Latin and Greek, and the massive collected letters of Huygens in French. She sometimes contributed original articles; her paper on *Homeric Astronomy* in 1887, where she discussed astronomical allusions in the *Iliad* and the *Odyssey*, was later to form a chapter in a book.

Agnes became also a regular contributor to the monthly astronomical magazine *The Observatory*, which catered for the British circle of professional and amateur astronomers. *The Observatory* (which still flourishes in exactly the same form) was founded in 1877 by William H.M. (later Sir William) Christie of the Royal Observatory, Greenwich, who was by this time head of that observatory with the traditional title of Astronomer Royal. This highly readable little magazine recorded the proceedings of the meetings of the Royal Astronomical Society, news items from home and abroad, reviews of books and letters from readers. The *History* had been favourably reviewed in its pages by the editor Edward Walter Maunder, who presumably invited Agnes' contributions, her only previous appearance there having been a modest note on early Italian observations of Saturn in 1883 when she was still unknown.[25] She was acquainted with a succession of editors over the years, and could always be relied on to provide items of current interest, which she did right up to the very week of her death.

Margaret Huggins

The publication of her *History* brought Agnes Clerke an intimate and cherished friend in Margaret Huggins (Figure 5.3), the wife and collaborator of the illustrious William Huggins. Huggins, one of the founders of astronomical spectroscopy, now in his late sixties, was the Grand Old Man of astrophysics. His story is well-known. A self-taught scientist without a university education, he had employed his modest private income to set up and maintain an observatory at his house in Tulse Hill, Clapham, then a leafy suburb in south-west London. He was one of the first to apply the spectroscope to the dim light of the stars, realising that here was an unexplored field of investigation, 'a spring of water in a dry and thirsty land'. His researches were immensely successful and epoch-making. He identified several chemical elements in the bright stars and discovered that many of the cloud-like hazy objects known as 'nebulae' are actually composed of hot incandescent gases. He also initiated the use of the Doppler shift in stellar spectra for measuring motions in the line-of-sight. In recognition of his achievements

Figure 5.3 Margaret Huggins.

the Royal Society had furnished his observatory with a set of instruments designed to his own specifications and built by the Dublin optical firm of Grubb. It was romantically told that it was through Howard Grubb, the head of that firm, that William Huggins first met his Irish-born wife Margaret, 25 years his junior.

Margaret Lindsay Murray was brought up in Monkstown, just outside Dublin. Both her parents were Scottish by birth. Her father was a Dublin solicitor and her grandfather a high ranking official in the Provincial Bank who had come to Ireland from Inverness in the 1830s. As a small child Margaret used to be taken outdoors in the evenings by her grandfather to be shown the constellations. She became an enthusiast for astronomy; she had a telescope with which she viewed sunspots and even managed to make herself a working spectroscope. She also took an interest in photography, then a fashionable pursuit of artistic ladies; indeed, she was an expert on the technical side, as her later work showed. By her own account she read widely about the advances of astronomy and was familiar with her future husband's pioneering work before she met him in person.

From the time of their marriage in 1875 the pair formed a highly successful scientific partnership, which continued for 30 years. Together they introduced the new technique of dry-plate photography to their observations, the first astronomers to do so. Though at first their work was published under Huggins' sole name, they later published under their joint names in the *Proceedings of the Royal Society*. When Agnes Clerke came on the scene, they were at the height of their powers.

Agnes Clerke and Margaret Murray had lived within a few miles of each other in Dublin in the 1860s, and again for several years in London, but had never before met. When the *History* came out, Margaret immediately recognised the author of the brilliant anonymous article on the 'Chemistry of the stars' published five years previously in the *Edinburgh Review*. 'Shortly afterwards', Margaret recorded, 'I entered upon a friendship and upon a companionship in astronomy which have been among my best pleasures.'[26]

Margaret Huggins, five years younger than Agnes Clerke, shared her love of music as well as of astronomy. She enjoyed taking part in Mrs Clerke's musical afternoon gatherings when Mrs Clerke herself, whose 'conversational powers were of a high order', delighted the company with her rendering of old Irish airs on the harp. She became a close friend of the family, though, unfortunately, never met the father who died in 1890.

The rapport between the two women had in it an element of the attraction of opposites. Margaret Huggins was vivacious, artistic and even a little bohemian. On the scientific side her talents lay in her practical skills as an observer and experimenter. Agnes Clerke was exceedingly shy, and happiest among her books. Her friendship with Margaret, and indirectly with the patriarch Huggins, was its own testimonial; in the course of time astronomers sought Agnes Clerke's ear as a route to Huggins' approbation.

Margaret Huggins was also a woman of strong personality and conviction. Though she liked to portray herself as merely her husband's assistant, it is likely that in reality she played a much more decisive role in their joint enterprises than she admitted. She would tolerate no criticism of her husband, either during his lifetime or after his death. The gentle Agnes Clerke, one feels, could be dominated by her friend; certainly the Hugginses could have found no better publicist for their researches.

Edward Pickering

With the American edition of her *History*, Agnes Clerke's name became well established in the United States. In that same year, 1886, Agnes Clerke began corresponding with Professor Edward C. Pickering of the famous Harvard College Observatory, who put her name on his mailing list.[27] 'Students of the Old World are deeply indebted to American investigators both for their brilliant results and for their generosity in communicating them', she wrote to Pickering on receiving a batch of publications from him. The first volume of the *Harvard Annals* received some months later she described as 'fresh proof of American originality and ingenuity'.[28]

This was a particularly important contact. In 1885 – the very year of publication of Agnes Clerke's *History* – Harvard College Observatory, already a centre for astronomical spectroscopy, embarked on what became one of the greatest ventures in that field. It was the photography of the spectra of thousands of stars obtained by means of a prism of small angle placed in front of the telescope lens, a so-called objective prism. In this arrangement, the image of each star in the field is spread out into a small spectrum, each photograph thus showing large numbers of small-scale spectra. The Harvard project was made possible by the Henry Draper Memorial, a liberal fund donated by the widow of Henry Draper, a distinguished astronomer and pioneer of astronomical photography. One of the fruits of that fund was the *Draper Memorial Catalogue* of stellar spectra, produced by the famous team of women under Mrs Williamina Fleming.

In June 1887 an exhibition of Harvard photographs was organised by Lockyer at the Royal Society in London. The Royal Society, with all-male fellowship, allowed women guests to be present at certain functions of a social kind. Agnes Clerke attended an evening soirée of that society for the first time on 8 June 1887 – undoubtedly at Lockyer's invitation – to see that exhibition. The same photographs were also shown by David Gill in the course of a lecture to the Royal Institution a few evenings earlier. 'Your stellar photographs have created a profound sensation here', Agnes Clerke wrote to Pickering, 'and there is and can be but one opinion as to their admirable character.'[29]

Agnes Clerke's unbounded enthusiasm for astronomical photography, which had suddenly become of prime interest among astronomers everywhere, was fired at this point. It found expression in an excellent essay, 'Sidereal photography', in the *Edinburgh Review* the following year, in which were discussed the Harvard work (Pickering's paper on stellar photography (1886) and his 'First annual report of the study of photographic spectra conducted at the Harvard College Observatory' (1887)), Gill's Royal Institution Lecture, 'The application of photography to astronomy' (1887), E. Mouchez's report on the Paris contribution to the photographic *Carte du Ciel* (1887) and Otto Struve's paper (in German) in the St Petersburg Academy of Sciences on 'Photography in the service of astronomy' (1886).[30]

6 A visit to South Africa

David Gill

The lecture on astronomical photography at the Royal Institution on 4 June 1887 saw the entry into Agnes Clerke's life of one who became a close friend as well as an important influence – David Gill, Her Majesty's Astronomer and Director of the Royal Observatory at the Cape of Good Hope, South Africa.

David Gill, one of the leading astronomers of the nineteenth century, was another of those who started his career as an amateur. Agnes Clerke described him perfectly as 'an astronomer by irresistible impulse' who, like Bessel [the great German mathematician and astronomer who first succeeded in measuring the distance to a star] 'exchanged lucrative mercantile pursuits for the scanty emoluments awaiting the votaries of the stars'.[1] Gill, a Scotsman from Aberdeen, studied mathematics and physics at Marischal College, University of Aberdeen, but left without completing his degree to take over the management of the family watch-making business. The necessity of regulating and setting his time-keepers led him to take a practical interest in astronomy and eventually to establish a public time-service for the city of Aberdeen – as Agnes Clerke's father had done on a smaller scale in Skibbereen. Astronomy soon became his all-absorbing passion, and when in 1872 the wealthy nobleman and astronomer Lord Lindsay (later Lord Crawford) decided to establish a fully modern observatory on his father's estate in Aberdeenshire, Gill leaped at the chance of joining him, despite the threat of insecurity and a much reduced income. It turned out to be a happy decision. The new Dun Echt Observatory, equipped by Lord Lindsay and Gill with the best instruments that Europe's top opticians could supply, was soon in operation.[2]

The first major undertaking of the Dun Echt Observatory was an expedition to observe the transit of Venus of 1874 from a site on the island of Mauritius. The purpose of observing this rare phenomenon was to determine the distance of the Earth from the Sun, a basic unit of measurement in astronomy and cosmology. During a transit, the planet is observed as a small black dot crossing the sun's bright disc; theoretically at least, the planet's precise position in space is calculable by comparing observations made simultaneously by astronomers in many stations on the face of the globe. Though the result of the huge international effort of 1874 did not live up to expectations, Gill was not prepared to give up. He determined to try another method, involving observations of Mars in 1877 when that planet was unusually close to the Earth. Gill staked everything on this venture. He gave up his salaried post and, with borrowed apparatus, he and his wife Isabella made their two-person expedition to the island of Ascension in the South Atlantic. Their endeavours were crowned with success, and in 1879 Gill was rewarded with an official appointment as Her Majesty's Astronomer with charge of the Royal Observatory at the Cape of Good Hope. In the next few years he earned honorary university doctorates, membership of various learned societies, Fellowship of the Royal Society and the gold medal of the Royal Astronomical Society.

The Cape Observatory was a sister observatory of the Royal Observatory at Greenwich. Its principal duty was to observe positions of stars in the southern hemisphere to complement those observed from the latitude of Britain. Though financially supported from Britain the enthusiastic and energetic Gill struck out on his own, improving and modernising the old-fashioned observatory he had inherited. Among his innovations was the use of photography as a means of mapping the sky. One photograph of a patch of sky, he pointed out, could reveal hundreds if not thousands of stars, far more than any observer could record visually one by one. He started a programme of sky photography with the aim of producing a star catalogue of southern stars, called the *Cape Photographic Durchmusterung* (CPD). He also enthusiastically supported an ambitious plan conceived by Admiral Ernest Mouchez, Director of the Paris Observatory, to have the entire sky photographed by world-wide collaboration, thereby creating a permanent record of every individual star down to a certain faintness.

The scheme bore fruit. In the spring of 1887 Gill attended the Astrographic Congress in Paris, where the ambitious international project of the photographic *Carte du Ciel* was launched, with Gill as president. After the congress Gill visited London and among other engagements delivered a lecture at the Royal Institution on the subject of 'The applications of photography to astronomy' which greatly impressed Agnes Clerke. The lecture coincided with a splendid exhibition of photographs from Harvard shown at the Royal Society and the Royal Astronomical Society. Agnes Clerke followed up these events with her excellent essay, 'Sidereal photography', already mentioned, in the *Edinburgh Review*, which is still quoted as a historical source.[3]

According to Gill's biographer George Forbes,[4] Mrs Gill was introduced to Agnes Clerke during their London visit – most probably by Mrs Huggins, an old friend – and was 'charmed by her artistic temperament'. She read the *History* and persuaded her husband to read it too – 'in spite of his belief that no woman could do justice to his noble science', surely a tongue-in-cheek observation on Gill's part. 'After he had read it through he was convinced of the intellectual power and originality of the authoress', wrote Forbes, and resolved to encourage her studies.

Agnes herself told a slightly different version of this story. Gill, seeing that the *History of Astronomy during the Nineteenth Century* was written by a lady, thought that it would be a nice introduction to astronomy for his wife and handed it to her with the remark: 'This, my dear, will probably suit you'. The lady, glancing through it returned it to her husband with a very emphatic 'I think it will suit *you*'.[5]

A close and enduring personal friendship grew up between the Gills and the Clerke family. Gill, who was a few months younger than Agnes Clerke, was a man of tremendous personality and warmth, and a voluminous letter writer. His lively correspondence contains lengthy accounts of his own activities, of his frustrations with his paymasters in London, discussions on astronomy in general, scientific gossip, and in Agnes Clerke's case, advice or elucidation on matters concerning her books and articles. Her letters to him – except those of a very personal nature – are fortunately preserved, and are the best source of information on the most active years of Agnes Clerke's life.

Gill declared that Agnes Clerke would profit from practical observational experience – the missing ingredient in her work – and proposed an extended visit to the Cape. 'I beg you seriously to consider the matter and try to arrange your plans and engagements as to allow you to leave London in August next and come out for a month or two to see us.'[6] 'You are not complete until you have seen and done a little practical astronomy. Your work would take on a new and higher character after a little practical knowledge. . . . For our sakes, for your own, and for the cause of astronomy I beg you to come.' 'It will do you the world of good and me a world of good also, just to have a real good talk about all the things you are in the midst of.' He promised her a nice quiet room all to herself for her literary work. 'Come you must, and right happy and welcome we shall make you!'.[7]

Observing at the Cape

Agnes Clerke did not find it easy to get away from her family or from her writing commitments. She explained to her American friend Edward Holden:

> Dr and Mrs Gill have most kindly asked me on a visit with a view to my gaining experience in observing. My last book, I know, showed the want of it; so I feel compelled to indulge myself in a great pleasure which I at the same time have to purchase by a strenuous effort. It is very difficult for me to break away from my writing engagements for so long as three or four months, to say nothing of the separation from my family, some of whom are far from robust. I have besides great doubts as to my own capacity for learning anything practical. I am not at home with instruments, and I am very short-sighted. So that I have every type of disqualification for observational astronomy.[8]

However, she succumbed to Gill's persuasions and on 9 August 1888, sailed from Southampton on the Union steamship *Mexican*.

As the ship moved southwards, 'crossing the line' on August 21, Agnes watched with pleasure the splendid southern constellations coming into view – the prominent stars of Centaurus, the southern cross reminding her of Dante's four stars, '*non vista mai fuor ch'alla prima gente*' (never seen before but by the first people), the bright knots of the Milky Way overhead, the planets Jupiter and Mars in Scorpio, and the faint zodiacal light. (This impressive rising into view of the southern sky was to be described in verse by Agnes' sister Ellen in her rendering of the tale of the Flying Dutchman, published in her anthology of poems.) The ship reached South Africa on August 30, the beginning of spring in that beautiful land.

Agnes Clerke spent two months at the Cape, living as the Gills' guest in their hospitable residence, which formed part of the Royal Observatory building (Figure 6.1). It was a time of particular excitement for Gill and his group of five or six devoted assistants. His two great astronomical projects were well in hand. One was the photographic catalogue of stars in the southern sky; the second was the measurement of the distances of minor planets, an extension of his earlier work on Mars.

Figure 6.1 The Royal Observatory at the Cape *c.*1890. Courtesy of the South African Astronomical Observatory.

Agnes Clerke was shown the observing routine involved in both projects, and was introduced, as promised in his earlier letter, to 'the pitfalls
and sources of error' in the computation of results from these 'delicate
observations'. She was also given a research project of her own with the
exciting chance of 'finding out a few new things which are all ready to be
found out, though I have never had time to seek them'.[9]

The project was in the field of stellar spectroscopy, with which
she was theoretically familiar but of which she had no practical experience. 'The spectroscopy of the Southern Heavens is absolutely virgin
soil', Gill had told her. 'A telescope with a direct vision prism on it, and
a selected list of objects, and time to examine the spectra of red and variable stars, should alone produce a crop of results.' The observations
were made with a 7-inch equatorial telescope (a very excellent instrument constructed by the famous opticians Merz of Munich) which had
a small prism attached to the eyepiece. With this arrangement one
could see a tiny rainbow or spectrum of whatever star was in view. The
scale of the spectrum was very small compared, for example, with what
the Hugginses could observe and photograph with their larger instruments. However, the sky could be quickly scanned for objects with
unusual spectra, and this in fact was what Agnes Clerke did.

The telescope was housed in a dome (which still stands) in the
garden behind the observatory, just a few steps from the Gills' living
quarters (Figure 6.2). Agnes Clerke became 'quite at home' with the
instrument, instructed by Gill and his secretary-cum-night-assistant,
Henry Sawerthal. One imagines Agnes, with her long skirt brushing
the none-too-clean floor, moving the telescope about in the darkness,
adjusting the various gadgets, consulting the star charts and making
notes in the light of a dim lantern. The work could be positively messy,
as when the declination circle fell off the telescope bringing with it a
pool of oil through which Agnes had to grope. But she took it all in her
stride, and Gill left the oily spots on the floor as a memento 'as sacred as
Rizzo's blood in Holyrood Palace'.[10]

In spite of a spell of cloudy weather Agnes succeeded in observing
up to twenty interesting objects, noting their spectra and colours.
These included the well-known luminous nebula Eta Argus (now
known as Eta Carinae, an HII region in modern terminology) and the

Figure 6.2 The dome where Agnes Clerke made her observations. Courtesy of the South African Astronomical Observatory.

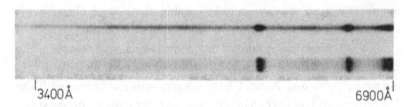

3400Å   6900Å

Figure 6.3 Spectrum of a star similar to Gamma Velorum (widened and unwidened objective prism photographs by UK Schmidt telescope) showing the bright emission lines.

beautiful emission-line star Gamma of the same constellation (now Gamma Velorum), the brightest star of its kind in the sky, the sight of which was 'something to be remembered' with its 'three bright rays, two golden, one of a cobalt hue', confirming earlier observations by Copeland and others (Figure 6.3). Agnes Clerke was also able to give an estimate of the magnitude of Eta Carinae (magnitude 7), which was known to be variable and to have recently brightened. Other objects were 'some fine specimens' of red stars with 'banded' spectra, not previ-

Figure 6.4 Agnes Clerke (centre) with the Gills and H. Sawerthal.

ously observed. Her results were published in two papers in the *Observatory* magazine.[11] She was in fact among the first to record the spectrum of Eta Carinae.[12] Looking back, one must regret that she had not time to extend her work into a thoroughgoing investigation of the southern Milky Way stars. This was sadly her only piece of independent research and ought by right to have been communicated to *Monthly Notices of the Royal Astronomical Society*.

By day Agnes was busy reading for her planned second book. Gill wrote to Ellen: 'Your sister sits opposite me in my study with a pile of books on either hand, which is gradually growing till she seems to be coming through a gate with rather badly built pillars on either side.'[13] (Figure 6.4)

The Gills were people after Agnes' own heart. With them, astronomy came first. It may not have been generally known at the time – though Agnes Clerke made it plain in an article she wrote on her return to London for the *Contemporary* magazine[14] – that Gill's photographic programme, which did not fall within his official remit, was carried out at his own expense (it cost him, in fact, £350 a year, half his salary). For the third time in her married life Mrs Gill had willingly made a sacrifice to enable her husband to follow his beloved vocation. 'The patines of bright gold with which Urania's treasure chests overflow are not of terrestrial coinage' was how Agnes Clerke poetically described the Gills' attitude to astronomical labours.

On the human side, the visit brought out the light-hearted element in Agnes' character. In his letter to Ellen, Gill reported that her sister 'was delighting everybody in the Observatory', and warned that when she returned home she would find her 'quite a Bohemian if not a "fair Barbarian"!'. The evenings, when weather was unfavourable, were spent in music-making. The Gills introduced her to their circle of friends at the Cape and presented her at Government House. One local newspaper in an article about Gill's work humorously called her the Queen of Sheba come north to visit Solomon!

Agnes also found time to look around Cape Town and its surroundings, admire the flora (she was interested in and knowledgeable about botany) and was able to contribute an account of the state of the Catholic Church in South Africa to the *Dublin Review*, the only article she ever wrote for that journal.[15] 'I carried away ineffaceable memories of kindness, and some treasured gain of astronomical experience', she wrote. 'Lessons of earnestness of purposes, stability of aim, and cheerful self-devotion can scarcely be missed by the itinerant lover of astronomy, in whose mind they will be tempered and illuminated by reminiscences of the beauty of the flowers by day and the glory of the stars by night'.[16]

Agnes Clerke sailed from the Cape on 31 October. In mid-ocean, 300 miles from the coast of Sierra Leone, the *Tartar*'s propeller broke down and the ship lay helpless for 36 hours in darkness. Rocket flares were sent up to alert passing ships, but, as Agnes Clerke humorously commented, 'ships at sea are like the police on shore, never found when most wanted'. A small French brig was sighted which offered to take the passengers on board but they declined, believing that it was engaged in native slave trade. Eventually the crew managed to reach Cape Verde to await another ship to England.

A friend for life

Agnes arrived home at the end of November. Her African adventure had been well worthwhile. Not only had she learned a great deal on the practical side; she also gained introductions to Gill's circle of international collaborators who would call on her in her home when they visited London. Gill was more convinced than ever of his pupil's abilities. He declared her to be 'one of the ablest women and most original of thinkers' that he had ever met.[17] Had he been on the council of the Royal Astronomical Society, he told her, he would have proposed her for honorary membership of the society, as bestowed in 1835 on two exceptional women of an earlier era, Caroline Herschel and Mary Somerville.[18] (She had just been elected an honorary member of the, admittedly less prestigious, Liverpool Astronomical Society.) He went further in a letter to Edward B. Knobel, secretary of the Royal Astronomical Society, asserting that 'a good deal could be said in favour of her claims' to the society's Gold Medal (which Caroline Herschel had been awarded in 1828) on the strength of her *History of Astronomy during the Nineteenth Century*; at least, he thought, she 'may be fairly entitled to the honour bestowed on Miss Caroline Herschel – that of honorary membership of the Society'. It would also have the practical value of giving her access to the library of the Royal Astronomical Society as of right instead of as a favour, especially as she was now preparing another book. Gill asked Knobel to ascertain the feelings of other members of the Council on the matter, but apparently without result.

Gill may not have felt able to do more at that time: the society's President was the Astronomer Royal, Sir William Christie, with whom Gill was on decidedly cool terms. It was to be many years before the suggestion to admit Agnes Clerke to the Royal Astronomical Society was acted upon, and she was never to receive any award from the Society.

As a result of her visit to the Cape, Agnes Clerke was on terms of close friendship with both Gill and his wife; and this friendship was enhanced through the Gills' periodic visits to Britain. The letters exchanged between them over two decades contain personal news and constant references to Mrs Gill's state of health. Mrs Gill was for much of her life a chronic sufferer of unclear ailments, and Agnes Clerke was

always warm in her sympathy with her. Gill always addressed Agnes
Clerke in his letters as 'My dear Friend'; she addressed him as Dr Gill
(or, after his knighthood, Sir David). As time went on their greetings
took on a more affectionate tone. Endings such as 'very truly yours'
became 'love to Mrs Gill and our cordial regards to yourself' and eventually (after ten years!) Agnes ventured to end her letter 'with our united
love to Mrs Gill and yourself (for why should we make a pedantic distinction?)'. Gill reciprocated and took to referring to his wife by name
(Bella), though Agnes Clerke, following the conventions of the times,
never addressed her as such.

The intimate nature of the friendship between the Gills and the
Clerkes showed itself on occasions, such as the death of Gill's sister and
brother-in-law in 1892, when the Gills decided to adopt their three
young orphaned nephews, aged 4, 7 and 8, whom Agnes had met during
the Gills' earlier visit.[19] Later, when Mrs Gill's sister planned a visit to
the Cape, the Clerkes arranged for her to stay with them when passing
through London. The Gills personal friends became their friends.

David Gill was without a doubt Agnes Clerke's closest and most
trusted friend outside her family, and a decisive influence on her thinking.

The Royal Observatory Greenwich

The Astronomer Royal's strained relations with Gill did not extend to
Agnes Clerke. Shortly after her return from South Africa, in the spring
of 1889, the possibility of a post on the staff of the Royal Observatory at
Greenwich was proposed to her in confidence.[20] The offer was probably
conveyed to her by Herbert Hall Turner, the energetic Chief Assistant
at Greenwich who did much of the administrative work of the Royal
Observatory and as editor of the *Observatory* magazine was a good
friend of hers. Her appointment, however, would have been an innovation – which is probably why the idea was kept confidential – as the
Royal Observatory employed only men at every level. Indeed, scientific
posts for women at that time were practically unheard-of.

The offer was especially tempting in that Agnes Clerke was

promised the use of a substantial telescope, the Lassell telescope,[21] for observing 'according to any plan I fancied'. That telescope, an equatorially mounted reflector of 2 feet in diameter, had been donated to the Royal Observatory Greenwich after the death in 1880 of its maker, William Lassell, by his daughter. Lassell, a wealthy engineer and amateur astronomer, was one of the great telescope makers of his age, ranking with his predecessors William Herschel and Lord Rosse in the construction of large speculum (i.e. metal) mirrors. With this 2-foot telescope he had discovered a satellite of Neptune in 1846, only a month after the discovery of the planet itself, Saturn's satellite Hyperion (simultaneously with Bond of Harvard) in 1848, and two moons of Uranus in 1851. The telescope was erected at Greenwich and kept in good order but was never used. In 1887 various additions had been made to it with a view to employing it for astronomical photography. The problem at Greenwich was one of manpower. The observatory had its official duties in positional astronomy, leaving in practice little time for the luxury of new research. The maintenance of the Lassell telescope was the responsibility of E.W. Maunder.

Agnes' response was that she would be glad to consider the offer when her current writing commitment (her second book) was out of the way. The free use of the telescope, as she told Gill, was a great attraction. Her formal duties, left vague, were to be 'literary'. On the negative side, the status of 'supernumerary computer', the rank at which she was to be recruited, was the lowliest in the observatory. The Astronomer Royal, however, was not to blame for this. His hands were tied by Civil Service rules as regarded the appointment of women, and staff numbers were in any case strictly controlled.

The staff of the Royal Observatory at that time consisted of the Astronomer Royal, a Chief Assistant, ten other astronomers (known as assistants), and a large number of young 'computers' trained to do routine computations and practical chores. Some of the assistants were university graduates; the older ones were men who entered at the age of about 18 by Civil Service examination. The computers, subdivided into permanent and 'supernumerary' (i.e. temporary), were men appointed straight from school at the age of 13 or 14, who often stayed only a few years before moving on to better jobs. By the nature of its work – meridian observations, time-keeping, reductions of star positions – the Royal Observatory operated largely as an assembly-line on which the computers carried out their set assignments.

In the matter of the employment of women, of whom Agnes Clerke was the first to be considered, the 'computer' budget was the only available fund. A nominal supernumerary computership with a proposed salary of £8 a month (roughly the salary of a secondary school mistress, and an improvement on the £6 maximum paid to other computers) was the best the Astronomer Royal could offer. The job, apparently specially created for her, was to be the first in an experiment to employ women at the observatory; Agnes Clerke's name heads a list of four women computers dated April 1890 preserved in the Greenwich archives.[22]

That list of April 1890 is the only reference in the Royal Observatory archives to Agnes Clerke's offer of employment. Before that date came round, however, Agnes had reflected on the practicalities of the case. The situation was very different from that at the Cape, where she lived as one of the family in the Director's comfortable residence and had only to step into the garden to reach her telescope. At Greenwich there was only one official house, occupied by the Astronomer Royal. Other members of staff lived wherever they found suitable accommodation. The observatory, on its hill within Greenwich Park, was reached on foot, and night observers were given a key to the main gate to let themselves in and out.

Greenwich was several miles from Agnes Clerke's home, so had she accepted the post she would have had to find somewhere locally to live. It was suggested that she could share accommodation with a young woman who was to be engaged at the same time. At the age of 47, Agnes Clerke would hardly have relished roughing it with someone half her age newly emerged from a Cambridge women's college. 'We could have kept each other company and perhaps even shared existence (if such it could be called) in lodgings', she remarked sarcastically to Gill afterwards. There were also, she reflected, 'almost insurmountable difficulties from the fact that Greenwich Park is said to be unsafe for ladies at night, so that a good deal of the glamour disappeared.' Though encouraged by the Hugginses with whom she discussed the matter, Agnes declined the job. 'Nevertheless I feel somewhat sore and sorry at having refused, and so shut out finally a prospect that was not without its attractions for me', she wrote to Gill in December 1889. As the Astronomer Royal's list of names carried the date April 1890, it is possible that Agnes Clerke gave some further thought to the matter before finally declining. After her refusal, the Lassell telescope was never put to use.

In practice, it was undoubtedly a wise decision. An observer as little experienced as she was, especially a none-too robust woman, would never have been able to handle an instrument as substantial as the Lassell telescope on her own. She would have had to rely on the goodwill and practical help of the already overworked male astronomers. One also wonders what programme of work she would have chosen. She might perhaps have considered Milky Way photography, though she had no direct experience in photography. Stellar spectroscopy, of which she got a taste at the Cape, would have involved acquiring a spectroscope: but in that case, she would have done better to join the already existing Greenwich spectroscopy programme of Christie and Maunder.

The women supernumerary computers – four in all – actually employed at Greenwich between 1890 and 1895 were products of the Cambridge women's colleges who had taken the mathematical Tripos.[23] Their terms of employment were the same as those of the young male computers and included night observing with the transit circle. So were their miserable salaries, which the Astronomer Royal was powerless to improve upon. 'I should think no educated woman would accept such a post at such a minute salary unless with a view to training for something higher, and getting into a mechanical routine would be fatal to that end', was Agnes Clerke's comment to Gill on the position of these women.[24] The 'lady computers' were seen at work for the first time by the Board of Visitors and other guests at the annual Visitation day in June 1890, though Agnes Clerke, who was invited, was unfortunately unable to be present to witness them. She was already on the Greenwich guest list for that function: in 1887 the elderly John Couch Adams of Cambridge, famous for his calculations predicting the position of the planet Neptune in 1845, noted in his diary that he attended a meeting of the Nautical Almanac Committee in Greenwich, and 'took in Miss Clerk' (sic) to lunch in the Octagon Room.[25]

Vassar College

Meanwhile, Edward Holden was anxious to recruit Agnes Clerke to an academic appointment in America. It was to the chair of Astronomy at Vassar College as successor to Maria Mitchell, its first holder and the first woman professor of any science in the English-speaking world.

Maria Mitchell was a women of very different temperament to Agnes Clerke. She was born and brought up in the whaling community in Nantucket, and was introduced to astronomy by her father, a man of many talents. He was a local observer for the United States Survey, for which he was provided with an alt-azimuth telescope; he also regulated chronometers for the whaling ships – rather as Agnes Clerke's father did for his local town. Maria, her father's helper from childhood, became a diligent student and practitioner of astronomy. In 1847, at the age of 29, she discovered a comet, which brought her to the notice of eminent American men of science. She was for many years employed as a computer of ephemerides (i.e. position tables) for the US Naval Observatory, and at the age of 44 was invited to take a chair of astronomy at the recently founded Vassar Women's College. Unlike the women's' colleges in England, which were attached to existing universities, Vassar was an independent, all-women, establishment, staff as well as students being women. Maria Mitchell retired from her chair in 1889, at the age of 70.

It was this post that Holden wished to secure for Agnes Clerke. It is worth recording Holden's warm recommendation, dated 6 August 1889, put to the Trustees of the College without consulting her.

> No woman has ever rendered the service to Astronomy that Miss Clerke has already given; and there are very few living men who have her philosophical grasp of the whole science and very few of equal erudition. It would be an honor to America should you invite her to your Chair of Astronomy.[26]

But Agnes responded to Holden:

> Circumstances sometimes oblige one to let slip occasions far beyond
> one's deserts. I am very doubtful whether I should be equal to it
> personally; but in any case I could not inflict the sorrow upon my
> parents of separating finally from them. They are advanced in life, and
> neither is strong. Indeed the wrench on both sides would be too
> severe.[27]

Holden's claims for Agnes Clerke were well justified, but so too was Agnes' assessment of herself as 'not equal' to university duties on account of her excessive shyness and her aversion to speaking in public. The chair was not, in the event, offered to her; it was filled by Mary Whitney, one of Maria Mitchell's favourite former students. The Trustees of Vassar wrote explaining their decision to Holden, who passed the letter to Agnes Clerke.

Back to writing

Agnes Clerke now concentrated on her second book, but without easing her topical contributions to the *Observatory* and *Nature*. Among these was an assessment of the possibilities of star-gauging by the latest, photographic, means.[28] Another was an account of the famous work of the Hugginses ('whose invaluable cooperation all lovers of astronomy must rejoice to see publicly recognised') on the spectrum of the Orion Nebula, presented by Huggins in a paper to the Royal Society in May 1889[29] – their first joint paper after 14 years of marriage. The picture revealed in Agnes Clerke's remark of the wife as an equal partner with her husband contradicted the general perception of her as the 'able assistant' (Huggins' own phrase) of the great man. It was an image abetted by his wife who devoted herself to maintaining her husband's almost legendary status by playing down her own part in their research. Agnes Clerke, who was on terms of close personal friendship with the Hugginses, would have known the true situation – confirmed by Barbara Becker's recent research into the Tulse Hill observing notebooks.[30] The notebooks, preserved at Whitlin

Observatory of Wellesley College, USA, show Margaret already in charge of observing plans in 1876, only six months after her marriage. The Edinburgh astronomer Charles Piazzi Smyth, who with his wife visited the Hugginses that same year, also gained the impression of her as the more modern half of the partnership:

> He has much liking for his old pieces of apparatus and for making up things himself, and was once above everyone else in spectroscopy. But now even with Mrs H.'s assistance he must see to it that others do not pass him in refinements and fine mechanics.[31]

The Hugginses' attempts to observe the Orion Nebula went back a long time. On 9 March 1882, Margaret Huggins described their first success in a letter to Robert Ball which made it plain that the work was jointly done:

> I cannot deny myself the pleasure of telling you that on Tuesday night, 7th, we succeeded in getting a photograph of the spectrum of the Great Nebula of Orion. My husband sent a paper giving results of examination of the photograph to the Royal Society yesterday. I was very busy helping him, or I should have sent you a line sooner, for I think you will be interested . . . What delights me very much is that our photograph is so far satisfactory that with longer exposure than we were able to give the other evening (owing to clouds coming up) we may, I think, hope for even better photographs of the spectra of other nebulae . . . Tuesday night was almost the only night during the last three years we had clear enough to give much success with a nebula, and even then we were hindered by clouds coming.[32]

The London skies could not compete with the clear air of the new mountain observatories. Several more years elapsed before their next observation of the nebula, published under their joint names.

Meanwhile it was no doubt satisfying to see *The History of Astronomy during the Nineteenth Century* translated into German. From some remarks in the correspondence with Gill it would seem that Agnes was not particularly happy with it at the early stages, but her wise friend advised her to check the proofs and not to worry so long as the meaning was understood.[33] One assumes that all went well in the end. The translation was published by Julius Springer, Berlin, and came out in early 1889.[34]

7 *The System of the Stars*

Planning and writing

Agnes Clerke's second book, *The System of the Stars*, appeared in November 1890, five years after the *History*. The author described it as 'a general survey of knowledge regarding our sidereal surroundings' – by which was meant the entire observable universe, believed to be confined within one single great agglomeration of stars. The first part of the meticulously planned book dealt with the characteristics of the various varieties of stars, the second with star groupings, nebulae, and, finally, with the structure and evolution of the cosmos. An appendix had useful tables, listing stars of different types, stellar motions and masses. The book was as successful as the *History* and was, moreover, more advanced and more sophisticated, being not merely a record of accepted facts but a critical discussion of their contemporary interpretation.

David Gill, who took an interest in the book from the beginning, remained involved in its progress throughout. As the chapters were written, they were sent to Gill who returned them with his 'scribblings in margins'.[1] Some chapters involved longer notes and warnings such as 'You cannot evolve star distances from your inner consciousness – you must be peculiarly careful about facts.'[2] He gave a caution about her uncompromising views on the finite nature of the stellar system. 'I cannot bear metaphysical questions – such as so many of my countrymen love – but I confess to you that I do not like the airy way in which you make the assertion "since the stellar system is of finite dimensions" ... If you say: "Provided that the stellar system is of finite dimensions" then so-and-so – you are then in a strictly logical satisfactory position'.[3] 'It is useless', he wrote, 'to pursue an argument on such a subject – you come at once to the unbreakable and insurmountable wall of the little hollow sphere which limits the mental conceptions of man, and which by death he alone can pass to the freedom of the space beyond – to the wider knowledge of God and his creatures.' But she clung to her position on the one-system universe.

It was a time of unusually rapid progress in astrophysics, and a great deal had happened in the five years since the *History* was published. One important development was the increased accuracy of observations of the Doppler shift in the spectra of stars, photographically recorded, from which their motions in the line-of-sight could be calculated. Leading the field was the German astronomer Hermann Carl Vogel, director of the recently founded (1888) Astrophysical Observatory at Potsdam, Germany, the first observatory in Europe to be devoted exclusively to research in astrophysics. Gill emphasised to her the importance of Vogel's work:

> Please learn, and inwardly digest Vogel's recent paper in the
> *Astronomische Nachrichten* on his photographic determination by
> means of his new spectroscope of stellar motions in the line of sight . . .
> I think it is the most important advance in practical astronomy made
> at one step for many a long day. His results in accuracy are to those of
> Greenwich as the accuracy of Bradley to that of Ptolemy.[4]

The reference was to Vogel's paper on the radial velocities of Rigel and Epsilon Orionis published in 1888, his earliest on that subject. The Greenwich measurements belonged to a long-standing programme of visual attempts to observe radial velocities initiated by Airy as far back as 1874, and continued for many years, mainly by Maunder. Lists of radial velocities of bright stars were published annually, but there were large discrepancies in the results. Greenwich was to abandon its radial velocity programme in 1890 in the light of the successes of the photographic method.[5] Obviously, Agnes Clerke's first draft, sent to Gill, had included an account of this work but, following Gill's advice, she excluded it entirely from the published book. Vogel's work was described, and his spectrum of Rigel reproduced, showing the H-gamma line displaced from the same line from a comparison source. As a result, Agnes Clerke had some correspondence with him on the subject[6] and was soon to make his personal acquaintance when he visited London.

Another epoch-making development while Agnes Clerke was writing her book was in the observation of eclipsing binary stars, a huge new field opened up by E.C. Pickering at Harvard. Following on the heels of the Harvard spectroscopists, Vogel in 1889 obtained impressive spectra of the best-known of these, the star Algol (Beta Persei). Algol, known to astronomers for centuries as periodically variable in brightness, was conjectured to be made up of two components, its brightness changes being explained in terms of eclipses of the larger one by a smaller companion. Vogel's spectra, revealing motions of approach and recession in step with the eclipse, was a remarkable confirmation of the double-star explanation. Agnes Clerke first heard of this exciting observation from Mrs Huggins in December 1889. 'Just heard from Mrs Huggins that Vogel's photos show Algol to have orbital motion 26¼ English miles per second', she wrote to Holden.[7] The case of Algol was found repeated in other instances of variables designated by Pickering as of the same type.

Also causing a considerable stir at this time was Norman Lockyer's so-called 'meteoritic theory' of the origin and evolution of stars and nebulae, heard for the first time in a paper to the Royal Society in November 1887 and published in full in *Nature* the same month. This theory postulated that all cosmic material was basically meteoritic, that is, composed of the same material as meteorites and comets. Support for this was supposed to come from the identification of stellar spectral features in the spectra of comets and of meteoric material in the laboratory. Comets, nebulae, bright-line stars, stars with banded spectra – all were supposed to be made up of meteorites at various temperatures. Immediately on its appearance Agnes Clerke wrote a five page report of Lockyer's paper in the *Observatory*[8] in which she explained Lockyer's arguments, realising their significance for future debate.

> The results communicated by Mr Norman Lockyer to the Royal
> Society are of a nature to modify profoundly current ideas as to the
> constitution of the universe. It is not possible to escape their influence.
> They may not all prove unimpeachable; the striking hypothesis by
> which they are rendered coherent may have to be rejected or modified;
> but opinion cannot stand where it stood before. A *col* has been crossed,
> and a new horizon widens out before us.

But – a problem not addressed by Lockyer – where did the meteorites come from in the first place? Her article ended:

> The small solid bodies which, more or less plentifully distributed, appear to pervade space, are in this theory treated as the fundamental atoms of the universe. But it is evident that we cannot begin there. They have a history, marked perhaps by strange vicissitudes. They may be agents of regeneration, but they are almost certainly products of destruction. Possibly they are seed as well as dust, and serve as the material link between the creation and decay of successive generations of suns.

Lockyer went on to enunciate a theory of stellar evolution based on his meteoritic hypothesis. The classification of stars by their spectra, which dated back to Secchi in the 1860s, was seen as a temperature sequence: white stars like Sirius were hotter than yellow stars like the sun, which were followed by red stars of various kinds. It was also seen as an evolutionary sequence, along which stars which began as hot objects moved as they became cooler and eventually faded away. Lockyer's scheme, enunciated in 1888, provided for a branch of ascending as well as of descending temperature. The stars began their careers as condensations from meteoric material, then became hotter, climbing to their maximum temperature before cooling off. It meant dividing stars of intermediate temperature into two groups, on the rising and the falling sides of the evolutionary graph. It was not easy to follow exactly how Lockyer divided the stars into such categories, and his views, which he never abandoned, were not accepted by the astronomical community at large.[9] However, historians recognise Lockyer's two-branch evolutionary theory as the first and only insight into stellar evolution to emerge before the 1920s.

Though unable to accept Lockyer's evolutionary theory in full, Agnes Clerke saw the merits of considering how stars might first have heated up before beginning to cool.

> Lockyer's classification of the heavenly bodies may, however, be considered independently of his meteoritic hypothesis. To a great extent it stands on its own footing; and it well deserves thinking about. Essentially an evolutionary scheme, it is the first of the kind with any

pretension to completeness. Stellar destinies are traced in it, so to
speak, from the cradle to the grave.[10]

Other spectroscopists, notably the Hugginses, had asserted that
observation did not support Lockyer's meteoritic theory. Soon after
Lockyer's paper appeared the Hugginses set out to test it, specifically
with regard to the spectrum of the Orion Nebula. In addition to the
lines of hydrogen, the spectrum of the Orion Nebula exhibited the mysterious nebular emission lines of which the strongest was the green line
at around 500 nm (500.7, identified much later as due to OIII).
According to Lockyer, this important line was due to magnesium
oxide, whose spectrum had a band head (what was then called 'fluting')
at this wavelength, magnesium itself being an important constituent of
meteorites.

The Hugginses put a great deal of effort into calibrating the wavelength scale of their spectroscope by matching up the spectrum of a
laboratory magnesium flame with the magnesium b band in the day-
light spectrum of the sun.[11] Their observations of the spectrum of the
Orion Nebula in the winter of 1888–89 satisfied them that the chief
nebular line most certainly did not coincide with Lockyer's supposed
meteoritic band. William Huggins announced this result in the
couple's first joint paper read before the Royal Society in May 1889.
Agnes Clerke unequivocally endorsed it in a detailed report in the
October number of *Observatory*.[12] 'Its upshot has been to establish an
unquestionable case of mistaken identity' [on the part of Lockyer], she
wrote. She had seen the evidence with her own eyes. '[How do] the
Hugginses refute Mr Lockyer's theory? Answer: With a Tulse Hill negative', she wrote in a letter to Gill.[13] In her book, she gave an excellent
exposition of the problem, explaining that Lockyer's wavelength coincidence, though near, was not exact. 'Spectroscopic agreements must
be absolute if they are to be reckoned significant.' She also rightly
pointed out that the 'magnesium fluting' had a diffuse edge while the
nebular line was sharp. She concluded: 'Nebular radiation cannot then,
it seems, be imitated in the laboratory. Possibly it signalises a modifica-
tion of matter arising only under extra-terrestrial conditions'. This
indeed turned out to be the true explanation; but this interesting idea

does not appear to have been taken up by anyone else. The mysterious lines were generally attributed to an unknown element which was to be given the name 'nebulium', a word invented by Agnes Clerke.[14]

Agnes Clerke thus spread news of the Hugginses' findings among her readers. The whole matter was soon to ignite an unseemly dispute between the Huggins and the Lockyer factions (from which Agnes Clerke, quite properly, remained aloof). Though her views on the nebular spectrum (which were correct) were to offend Lockyer's young assistants, who took the whole debate personally, they did not impair her friendship with Lockyer himself. With perfect tact she wrote to congratulate him on being awarded the French Janssen Prize for his work on the sun,[15] and soon afterwards (January 1890) visited him at his observatory before he set out for Egypt to pursue his latest interest, the astronomical aspects of the ancient temples.[16] She found him 'well, and in great spirits about meteorites and everything else'. Later that year (11 October 1890), as her book was about to come out, she found herself next to him at dinner one evening.

> Though painfully sensible of some mutual constraint on the burning topic of meteorites, we get on quite well. He tells me he plans to go out again to Egypt in January with a view to measurements and orientations of Temples. Already he believes he has nearly succeeded in reconstruction of the old Egyptian constellations and intends to check the chronology . . . by the stars.[17]

His book *The Meteoritic Hypothesis* had just been published. '[It] embodies and brings together all the scattered papers in which his recent researches were described, and so brings them in a more available form before the scientific world. Although the verdict of the Lick spectroscope has not been favourable to their results, they have had the good effect of stimulating enquiry, and defining the lines it should set along', wrote Agnes Clerke to Holden.[18] The reference was to recent observations by James Keeler at Lick of the spectrum of the Orion Nebula which corroborated the Hugginses' (as against Lockyer's) interpretation but had come too late to find a place in her book. Agnes Clerke's measured assessment of Lockyer's work contrasts with that of the 'exultant' Huggins who, on hearing Keeler's result by telegram, had written a crowing letter to the *Times*.[19]

The earlier of these conversations with Lockyer was at one of his observatory social evenings. The company included Isaac Roberts, whom Agnes very much liked, a wealthy amateur astronomer with an observatory near Liverpool who had taken part in the Paris conference on astronomical photography with Gill the previous year. He showed her some of his original negatives which she could examine with a magnifying lens. (The magnifying lens was Agnes Clerke's only scientific instrument.) She was especially impressed by the structural details of the globular cluster in Hercules, M13; the apparent lanes among the stars, originally noted by Lord Rosse, and a 'perfectly symmetrical ring of stars', were pointed out to her by Roberts. (The lanes were in fact illusory, and disappeared on long-exposure photographs.) She made a sketch of the cluster with its lanes and sent it to Holden,[20] and in her book contrasted the appearance of M13 with her own observations at the Cape of the great southern globulars Omega Centauri and 47 Tucanae. Roberts also showed her his double exposure photograph of the Orion Nebula on which every star was double – a method of detecting variables from the inequality of the dots.

Also present at Lockyer's party were W.H. Guy, a physician and expert in medical statistics, now well on in years, 'but unaccountably lively and even jocose' and 'Professor [William E.] Ayrton [the electrical engineer], obviously Scotch and therefore dear to me on the spot, and a young lady who photographs etc. for Mr Lockyer (women are coming to be found to be very convenient) and her chaperone of the nondescript character of that class.'[21] Ayrton's wife Hertha (née Marks), an extremely talented physicist, would have the distinction of being awarded the Hughes medal of the Royal Society in 1906, though a proposal for Fellowhip of the Society was turned down. Lockyer was an advocate of women's equality. Agnes Clerke's remark concerning Lockyer's female assistant underlines the restrictions on women's entry to the scientific world, when it would have been unacceptable for a woman to work alone among male associates. Presumably Lockyer had to pay the companion as well as the assistant.

Roberts was about to move his observatory to what he hoped was a more favourable location at Crowborough in Sussex. At Crowborough he replaced his 15-inch silvered glass reflector with a 20-inch one,

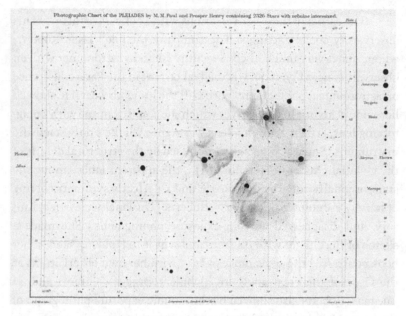

Figure 7.1 Map of the Pleiades, photographed by the Henry brothers.

having donated the former to Robert Ball to be installed at Dunsink Observatory of Trinity College Dublin. Four of Robert's photographs of nebulae and that of M13, as well as one of the Pleiades by the brothers Paul and Prosper Henry of Paris (Figure 7.1), embellished Agnes' book. The Henrys, whose beautiful sky photographs had been the original inspiration for the international *Carte du Ciel*, were the designers and builders of most of the special astrographs (photographic telescopes) provided to observatories world-wide for the project.

On the technical side, an exciting new development at this time was the photography of Milky Way panoramas by Edward Emerson Barnard at Lick Observatory. In July 1889 Agnes Clerke's enthusiastic adviser Edward Holden sent her what he called 'an exquisite positive copy' of a photograph which Barnard had taken the previous night, 'a lovely picture which I ask you to accept with Mr Barnard's compliments and my own'.[22] Barnard's venture into sky photography had been prompted by his photographs of the solar corona, taken during the total eclipse of 1 January 1889, observed in California. He had used a wide-angle portrait lens and was amazed at the amount of fine detail revealed, far more than could be seen on exposures taken with a large conventional telescope. He tried the same camera on the starry sky, and obtained beautiful images of star-clouds and nebulosities of a kind never before recorded.

Holden recommended viewing Barnard's photograph by holding it up in front of a soft white porcelain shade of a kerosene lamp when it would look well even under an eyepiece of 1.5-inch focus. It excited Agnes Clerke's 'unbounded admiration'.[23] She was the first astronomer in Britain (and probably in Europe) to see examples of Barnard's pioneering Milky Way work. Two of the photographs were reproduced as woodcuts – which unfortunately did not do them justice – in her book.

A month later (August 1889), in an article in the *Publications of the Astronomical Society of the Pacific*, Holden gave great offence to Barnard by describing the Milky Way photographs as 'experiments'. Barnard insisted on publishing his own account in the next issue of the journal, pointing out that his photographs were an innovation, 'the only photographs ever made, here or elsewhere, which show all the true Milky Way'.[24] Holden had used the same expression, 'experiments', in a letter to Agnes Clerke which accompanied a parcel containing two more glass positives of Barnard's photographs, one for herself and the other, a field in Sagittarius (Figure 7.2), to be passed on to Huggins 'with my compliments and kindest regards. He may care to make copies of the others which you have. They are very interesting as experiments, tho' they have their faults, which he will see as soon as anyone. They were exposed by Mr Barnard'.[25] The quarrel between Holden and Barnard over these photographs was the first glimpse by the outer world of the tensions brewing between the Director of Lick Observatory and his staff. Nevertheless, Barnard was kind enough to write some months later to Agnes Clerke at Holden's request, helping her to identify various objects.[26]

Holden sent some Lick photographs to the Royal Astronomical Society, which were exhibited at the November 1889 meeting. Agnes Clerke, who could visit the Society rooms freely (but as a woman was not yet permitted to attend meetings), was shown them in advance by the Society's secretary, her friend and fellow music lover, E.B. Knobel.

Figure 7.2 Milky Way field in Sagittarius, by E.E. Barnard.

William Wesley, the administrative secretary, who for a whole generation was regarded as an expert on the interpretation of astronomical photographs, pointed out details on the photograph of the Andromeda nebula to her.[27]

While writing her book, Holden himself had showered Agnes Clerke with material, mainly of a general nature, about activities at Lick – a *Guide to Lick Observatory* (1888), his *Address to the Astronomical Society of the Pacific* (1889) and popular articles in magazines, interspersed with discussions on literary matters. On one occasion he even sent her a transcript of his observing notes. To a query about the structure of the Milky Way, he responded 'in a portentiously long letter which you must lay to my pleasure in speaking to you, who appreciate

so thoroughly'[28] explaining his plan for a star-counting programme
which would examine the increase in numbers of stars with magnitude.
The correspondence between Agnes Clerke and Holden, with their
almost romantic interest in astronomy, is quite moving – she bringing
up topics of interest, he responding warmly and loquaciously. However,
Gill hinted to Agnes that Holden was 'attempting too many different
things with the Lick telescope' and after what Vogel had done, should go
straight into stellar spectroscopy by photography.[29] Agnes herself realised that there were 'too many irons in the fire there already',[30] but kept
the Lick work in the eye of the astronomical community in Britain with
an informative article about it in the *Observatory*.[31]

All the material for her book, collated and shaped into a systematic account of the Milky Way galaxy and its contents, was far more
than the promised 'general survey of knowledge regarding our sidereal
surroundings'; it was a triumphant demonstration of the one-system
model of the universe, a model not finally overthrown until the 1930s.
The final chapter appropriately borrowed its title from William
Herschel, 'The construction of the heavens'.

The one-island universe

Agnes Clerke's dictum in *The System of the Stars* is often quoted as representing the opinion of the majority of astronomers of her time[32]:

> No competent thinker, with the whole of the available evidence before
> him, can now, it is safe to say, maintain any single nebula to be a star
> system of co-ordinate rank with the Milky Way. A practical certainty
> has been attained that the entire contents, stellar and nebular, of the
> sphere belong to one mighty aggregation, and stand in ordered mutual
> relations within the limits of one all embracing scheme.[33]

The crux of the matter was the status of the nebulae. Stars and
star clusters favoured the belt of the Milky Way and clearly belonged to
a single system. Irregular nebulae (gaseous and planetary nebulae),
interspersed among the stars, were also certainly members. The so-
called 'white nebulae' (eventually recognised as galaxies) were quite

differently distributed, populating the polar regions of the Milky Way and being absent from the belt of stars. (The true explanation of this 'zone of nebular dispersion' – interstellar absorption in the galactic plane – had not yet emerged). The Herschels, father and son, had noticed the crowding of nebulae in certain parts of the sky and their absence from others; but it was left to R.A. Proctor in 1869 to confirm the differing distributions of the two types of object by a proper statistical examination of their positions in the sky. These complementary distributions were interpreted as 'the subordination of stars and nebulae alike to a single idea embodied in a single scheme.'[34] Agnes Clerke called in support on this point the English philosopher Herbert Spencer, who busied himself with, among other matters, cosmological evolution. He had declared (in 1854) that the evidence for a physical connection between stars and nebulae was 'overwhelming'.

'The grandiose but misleading notion that the nebulae are systems of equal rank with the galaxy', wrote Agnes Clerke, had to be dismissed. There is but one 'island universe', she declared, to which all objects, of every kind, belonged. A convincing illustration was the apparent co-existence in the Magellanic Clouds of stars, globular clusters and nebulae of every description. The Magellanic Clouds, now known to be close satellites of our Galaxy, are visible to the naked eye and look rather like detached pieces of Milky Way. John Herschel, in his survey of the southern skies in the 1830s, had noted their constituents, though in reality there are no spiral nebulae in the Magellanic Clouds. The Milky Way system, embracing everything, was seen as having a complex structure, with ourselves in a 'nearly central position'.[35]

In the case of the greatest of the nebulae, the Andromeda Nebula, an apparently unanswerable proof that it could not be an external galaxy was at hand. To assume that the nebula had the same dimensions as our Milky Way would mean placing it at the huge distance of over 300,000 light years away. This meant that the star that suddenly appeared there in 1885 (Nova Andromedae) would have had to be equivalent in brightness to 50 million suns – an utterly impossible result, according to the beliefs of the day. The object was in fact a supernova, and was indeed unimaginably luminous.

There was another potent argument against the existence of

multitudes of external galaxies. This was Olbers' paradox, though Agnes Clerke did not call it by that name. It stated that, if the universe was infinite and filled with stars, the night sky ought to be luminous all over. In her words: 'From innumerable stars a limitless sum-total of radiations should be derived, by which darkness would be banished from our skies'[36] – unless, that is, there were some sort of absorption of light in interstellar space, for which, she asserted, there was no evidence.

The distances of the nebulae were as yet unknown. They had no discernible motions, which proved that they were certainly at greater distances than the nearer stars. But, in Agnes Clerke's judgement, this did not place them outside the boundaries of the Milky Way. They were probably at 'the same order of remoteness' as eleventh or twelfth magnitude stars,[37] with motions 'of an extremely sluggish nature'[38] which better observations, expected from photography, would eventually reveal. But what was the actual nature of the high-latitude nebulae? Were they remote unresolved clusters of stars, vastly more distant than the equatorial ones, or were they composed of smaller constituents, perhaps even of 'star-dust'? Agnes Clerke preferred the latter picture. 'While galactic nebulae are of what we may roughly describe as stellar composition, non-galactic nebulae are more or less pulverulent'.[39] Though not claiming to know all the answers, it seemed the case that stars originated in nebulae, the transition from nebula to star being so gradual that we see contemporaneously in different types of object the various stages of this evolution. The entire system was assumed to be in a state of rotation – otherwise it would collapse under gravitation – though the details of this, and the reasons why the nebulae should tend towards the poles while the stars gathered in the equatorial plane, were problems which would 'only be investigated by long and arduous methods'[40] (or, as even more vaguely expressed in the second edition of the book, left 'to the dim futurity').[41]

The book published

In the final stages of writing, with new observations coming out all the time, Agnes Clerke 'felt like one of the Danaides at work with her tub.

As fast as my chapters are re-written and off my mind, they get superannuated'.[42] She arranged that the printing of the Appendix should be left to the last moment to allow possible new pieces of information to be added. Even so, by the time the book came out, she felt that there were many passages already in need of updating.

A glance at the Preface to the book illustrates how far Agnes Clerke had advanced since she had first appeared on the scene. While two names only, those of Holden and Copeland, were acknowledged for their help in the preface to her *History*, there now were more than a dozen, who included the front-line astrophysicists at Lick, Harvard, and Potsdam, as well as the Hugginses and David Gill. Gill was singled out for having given her the opportunity of observing the southern skies, and for reading many of the chapters in advance of publication.

The manuscript was finished on Easter Sunday (6 April) 1890. The final stages of writing were sadly overshadowed for Agnes Clerke by the death of her adored father and first teacher. He died after two months' illness on 24 February 1890, aged 76. 'It is with a heavy heart that I am now working on at the book he would so keenly have enjoyed the publication of', she wrote to Holden,[43] 'but no doubt it will be made all right for us in a better way than the one we would choose.' She dedicated the book to his memory. The title page carried a motto from Dante:

> Io vidi delle cose belle
> Che porta 'l ciel.
> ('I saw the fair things that Heaven holds')

from the last lines of the *Inferno*, when the poet emerges from the darkness and sees the sky once more.

The System of the Stars was published by Longman's, publishers of the *Edinburgh Review*. A. and C. Black, who had published her *History*, had wished her to write instead a book for a more popular readership. As her second book went to the printers Agnes Clerke had a visit from Adam Black, 'a cautious and canny sort, but with a reality about him which we could not help liking', whose firm was now moving from Edinburgh to London. Black repeated the offer made earlier for a manual on elementary astronomy as part of a popular science series.

She had 'a tendency in me to do what I am asked', but on her brother's resolute advice she refused: the time was too short and the profits too small, he said. 'I breathe more freely to be out of it', she told Gill.[44]

A Baltic cruise

Agnes Clerke was never robust: the writing of *The System of the Stars*, on top of other literary commitments, combined with the loss of her father, left her exhausted. Her solicitous friends Henry Reeve, editor of the *Edinburgh Review*, and his wife brought her to stay at Foxholes, their charming home on the Hampshire coast, and arranged for her to join a cruise in the Baltic aboard the large and splendidly equipped yacht *Palatine*, as the guest of their friend Mrs Watling, 'a charming person and full of kindness'.[45] Agnes Clerke hoped to combine the holiday with visits to observatories in Scandinavia and Russia.[46] She was warmly welcomed at the observatories of Copenhagen and Stockholm by C.F.Pechüle and Karl Bohlin respectively. Pechüle, during a Transit of Venus expedition on the island of St Croix in 1882, had observed southern emission-line stars including Gamma Velorum, which Agnes Clerke had seen for herself from the Cape. Bohlin, then a young man, became known later for work on the structure of the Milky Way.

The highlight of her two-month tour as an 'itinerant astronomer' was to be a visit to the great observatory at Pulkovo near St Petersburg. Her good friend H.H. Turner had provided her with a letter of introduction to Hermann Struve, astronomer son of Otto Struve, the recently retired director. She sent the letter up to the observatory on arriving in St Petersburg but as there was no response she was too polite to make herself known without the formalities. She could hardly have chosen a less auspicious time for her visit. Otto Struve's successor, Fedor Bredechin, director of the Moscow Observatory, had taken over in the spring of 1890,[47] greatly to the dismay of Struve who had hoped that his son, who was on his staff, would have succeeded him, as he had succeeded his own father. Agnes Clerke was unaware of all this, and of the fact that the younger Struve had already left Pulkovo to continue his career in Germany. It must have been deeply disappointing for her not to have seen the institution which in her *History* she had described as 'surpassing all others of its kind in splendour, efficiency and completeness'. In spite of her disappointment, however, she deemed the Baltic cruise an important event in her life; it is mentioned, together with her journey to South Africa, in her *Who's Who* entry. And no doubt her health was restored.

Reviews of *The System of the Stars*

The System of the Stars came out in November 1890. Before even cutting the pages of her own personal copy, Agnes Clerke sent a copy to Gill.

> I only wish it could carry with it all the sentiments of gratitude and regard which the remembrance of your assistance in its composition can never fail to excite in me ... I am sure now that you will add still further to my debt of obligation by jotting down corrections ..., no doubt they will occur to you plentifully, and they will all be valuable to me. Already I could myself wish some things in it different; but my long absence while it was going through the press made the correction of the proofs unsatisfactory.[48]

Another copy of the book went to Holden, who reacted in his customary (to Agnes Clerke at least) effusive manner.

> I have pretty well read your *System of the Stars* through – and I wish to give you my hearty congratulations upon it. It is not only an adequate account of our present knowledge but it is full of pregnant suggestion for our future guidance ... The highest quality of the work is its philosophical bent – tendency – and you have not lost sight of general principles in the midst of all the tangle of detail. It seems to me that in spite of the hopelessness of the subject, so to speak, you must be satisfied – not only satisfied that you have done your best, but satisfied that you have made a permanently valuable addition to the thought of our generation. .... I am pretty sure there is no English-speaking astronomer that could have accomplished more, taking the book as a rounded whole. I feel for you, personally, that you could not present your finished work to your father – but it is a splendid tribute that you offer to his memory.[49]

Several reviews appeared in the scientific journals. One, seven pages long, by Margaret Huggins, in the December number of the *Observatory* magazine[50] was largely a eulogy of her friend, written in her distinctive flowery style with many literary allusions. She well summed up the essential nature of the book compared with the usual popularisations. '*The System of the Stars* will not minister to people who are too lazy to do more than attend scientific lectures to have their ears tickled, but will minister to those who long to know, and who do not grudge the best intellectual exertion they are capable of', and (referring to the international spirit of the book) 'we are not humiliated by any attempt to make Great Britain the hub of the astronomical world.'

A more important review in *Nature*,[51] signed 'F', was by Alfred Fowler, Lockyer's young assistant at the Solar Physics Observatory at South Kensington whom she already knew. He praised the book on the whole, but found fault with the doubts expressed on his master's views on the meteoritic origins of the stellar world. Holden's warm letter of praise, as she wrote to him, came 'with especially consolatory effect after a discouraging review in *Nature*, pointing out, fairly enough, what I know perfectly to be weak points in the book. I am indeed always glad to be really criticised; but correction comes more pleasantly from a sympathetic authority, such as yourself, should you at any time tell me, publicly or privately, what you consider faulty or erroneous.'[52]

Holden gave the book for review in the *Publications of the Astronomical Society of the Pacific* to George Ellery Hale, who devoted no fewer than fifteen pages to it.[53] Hale, who like Fowler was only 22 years of age, was a rising star of American astronomy, already the inventor of the spectroheliograph; in fact his first positive results were achieved in the very same week (April 1890) in which Agnes Clerke had finished the manuscript of her book. That summer, newly graduated and newly married, Hale with his bride made a special trip to Lick Observatory on Mount Hamilton, his first experience of a mountain site. He met for the first time the Director, Holden, who offered him the use of the 36-inch telescope to try out his spectroheliograph.[54] It was Agnes Clerke's good fortune that Holden should have asked his guest to review her book and thus introduce her to a brilliant astronomer of the younger generation.

In his review, Hale concentrated on spectroscopy, going into the minute details of identification of spectral classes, line strengths, indications of stellar evolution, etc. He did not hesitate to chide the author for jumping too easily to conclusions regarding complex matters (she was always apt to be enthusiastic).

> It seems best to defer judgement until greater certainty is secured ... It cannot fail to be noted that the majority of the best spectroscopists are extremely conservative, and few of them hazard a general explanation of the complicated problem whose great extent they so fully realise.

However, he took her part in the debate with Fowler, whose review had preceded his. In the main, his opinion was favourable and encouraging. Agnes Clerke took his remarks seriously and resolved to learn from them.

None of these reviewers commented on the cosmological aspect of *The System of the Stars*, their main interest being the natures of individual objects and groups of objects. An article in the widely read popular periodical, the *Athenaeum*, however, showed that the universe depicted by Agnes Clerke was not an entirely familiar one to the non-scientist. 'It is perhaps even now not so well known to the general reader as it should be', wrote the anonymous writer, 'that the theory (first apparently put into shape by Wright in 1750 and for a considerable time held strongly) that the nebulae are as a class at immeasurably greater distances than the bodies that appear to us as stars, forming as it were separate galaxies, has long since been abandoned by astronomers.' But he went on to caution: 'Nevertheless such extensive galaxies may undoubtedly exist; for we know far too little of the nature of the luminiferous ether to be able to say with any confidence how far our sense of sight may carry us into space'.[55]

The mantle of Mary Somerville

Just as the *History of Astronomy during the Nineteenth Century* was a sequel to Robert Grant's History of 1852, so *System of the Stars* may be fairly regarded as filling the place once occupied by Mary Somerville's

Connexion of the Physical Sciences, first published in 1835 but still in print after going through numerous editions. *Connexion of the Physical Sciences* was a non-mathematical account of the world addressed to an intelligent lay readership. It covered, in addition to astronomy, properties of matter, optics, heat, electricity and magnetism; and demonstrated how all these sciences were connected through the universal laws of physics. The solar system in all its intricate detail was subject everywhere to these same laws. Mary Somerville's *Connexion of the Physical Sciences* had an appreciative readership among physicists and astronomers. James Clerk Maxwell, genius of the new generation of physicists, described it as one of those books 'which put into definite, intelligible and communicable form, the guiding ideas that are already working in the minds of men of science, so as to lead them to discoveries, but which they cannot yet shape into a definite statement.'[56]

Though Agnes Clerke did not set out deliberately to imitate Mary Somerville, her *System of the Stars* was in many ways a 'new astronomy' version of *Connexion of the Physical Sciences*, the laws of physics now extending all the way to the stars. Both books give pictures of the physical universe as it was known at their respective eras. Both were products of mature minds – Mary Somerville and Agnes Clerke were respectively 54 and 48 years of age when they wrote them – and were results of careful study of original sources. And though Agnes Clerke did not have her predecessor's mathematical genius, she was well equal to her as an expositor. The mantle of Mary Somerville fitted her well.

8 Social life in scientific circles

The British Astronomical Association

Apart from the Royal Institution, which welcomed the public to its scientific lectures, there was for a long time little opportunity in London for Agnes Clerke and like-minded women to foregather with people who shared their enthusiasm for astronomy.

Men astronomers, amateur and professional, enjoyed a common fraternity at the Royal Astronomical Society, founded in 1832, which met every month in London. A fraternity it truly was, as women were excluded by statute from its ranks. Fellowship of the Society was open to any man with an interest in astronomy, provided he was duly nominated and paid his fees. It was at the same time much more than a social club. The Presidency of the Society or the award of its gold medal were high accolades, and many leading members were also fellows the Royal Society.

In 1890 the British Astronomical Association was founded to cater for the interests of amateur astronomers, some of whom were dissatisfied with the Royal Astronomical Society because the fees were high and because women were ineligible. Some complained also that the society was becoming too academic. The suggestion of forming a society to include ordinary lovers of astronomy came originally from W.H.S. Monck, the Dublin amateur astronomer who had been Aubrey Clerke's student contemporary, in a letter to the *Observatory* magazine, citing as a model the older Liverpool Astronomical Society to which Agnes Clerke belonged by correspondence since 1885. The matter was taken up by E.W. Maunder, who advertised the new society in the popular magazine *English Mechanic*. There was an overwhelming response; over 250 enquiries were received within a few weeks.[1]

The Liverpool Astronomical Society founded in 1882 was a lively and well-run organisation where observers were divided into groups or sections, each with its own project. Reports were sent in by members as well as read out at meetings, and were published in their *Observations of Proceedings*, renamed *The Journal of the Liverpool Astronomical Society* in 1884. Particular interests were observations of the sun and of aurorae. The Director of the solar section was the remarkable Miss Elizabeth Brown, the only woman in Britain to have her own properly equipped observatory.[2] To these activities were added in 1886, at the request of Urban Leverrier of Paris, a search for the hypothetical infra-Mercurial planet Vulcan, which would be expected to show up as a spot on the sun's disk. (Vulcan was conjectured to explain the discrepancy from theory in the motion of Mercury; this discrepancy was eventually explained by the theory of relativity.) Such a search had been carried out, without success, on the previous favourable date in 1877 by Father Stephen Perry, a Jesuit astronomer at Stonyhurst College in Lancashire who co-ordinated the Liverpool effort.

Miss Brown's sunspot work had put her in touch with Maunder at Greenwich, and she was thus an adviser to as well as a founder member of the new Association. The inaugural meeting of the Association took place on 24 October 1890 – a few days before the first copies of *System of the Stars* were delivered – and was attended by Agnes Clerke and her brother Aubrey, who both joined. Their sister Ellen joined later. The Hugginses also supported the new venture: Huggins was elected a vice-president while his wife and Agnes Clerke were among four women elected to the 48 strong council.[3] Though Agnes Clerke enjoyed scientific company and scientific conversation, she did not seek the lime-light nor did she have much interest in a public role for herself. She told Gill 'I have been unfittingly elected a member of the Council. I am totally useless but must attend a few of the meetings for form's sake.'[4] She remained on the council for three years and was re-elected to it in the session of 1896, doing her duty as head of the Spectroscopy section according to the Association's pattern of organising its activities by topic. However, she does not figure prominently in its published proceedings: the fact is that she had little to gain from it. The Hugginses, too, while patronising the association from on high, took care to send their scientific papers to the Royal Society.

The foundation of the British Astronomical Association effectively marked the separation of professionals from hobbyists in British astronomy (independent self-supporting astronomers like Huggins or Monck probably regarded themselves as professionals). It may not have appeared so to the enthusiastic members who shared the first meeting with the President Arthur E. Downing, Lord Rosse and William Huggins, all Fellows of the Royal Society, not to mention several Fellows of the Royal Astronomical Society. Turner, Chief Assistant at Greenwich, informed Lockyer that the new society had not been formed 'aggressively' (i.e. towards the Royal Astronomical Society) 'but, as a professional astronomer I shall have nothing to do with the new one which I think is really meant for amateurs.'[5]

Professional scientists, meaning those officially employed and salaried, were at this time increasingly filling posts in universities and colleges in London and the provinces as the gentlemen amateurs of the nineteenth century became obsolete. The campaign that hastened the general professionalisation of science was due in great measure to the dynamism of the biologist Thomas Huxley and his circle, in the wake of the Darwinian revolution in biology. Norman Lockyer, a long-time friend of Huxley's, was also dedicated to this cause: having started off financing himself he was since 1875 on the staff of the Government-supported Royal College of Science in London. The number of salaried posts in astronomy in Britain was small – reckoned in tens rather than in hundreds among 600 Fellows – so that the Royal Astronomical Society never became exclusively 'professional'. Indeed, one of its attractions has always been its mixture of amateur and professional astronomers.

Foreign amateur societies

Amateur societies were the fashion at the end of the century. The Astronomical Society of the Pacific in California, which pre-dated the British Astronomical Association, was founded, as already described, by Edward Holden in February 1889 with Holden himself as President. Holden proposed Agnes Clerke for membership at an early stage and included her *History of Astronomy during the Nineteenth Century* in the society's recommended reading list – a helpful advertisement at the time. She declared to Holden that her election in 1889, the first non-American member, 'a signal honour': the society's 'invigorating activity', she said, made England feel a 'Sleepy Hollow'[6]. In return she contributed an article on double stars for the Society's journal, *Publications of the Astronomical Society of the Pacific*.[7]

The year 1890 also saw the founding of the Astronomical and Physical Society of Canada in Toronto.[8] There were few active astronomers in Canada at that time, and in its early days the society relied entirely on papers contributed by non-local members and read out at their meetings. Established astronomers from abroad were made honorary members while others could become corresponding members. Agnes Clerke joined in the latter class, donated a copy of her *History* and sent a paper on the recent nova, Nova Aurigae, which was read out at the Society on 12 July 1892 and later published in the *Observatory*.[9]

The Cardiff meeting of British Association for the Advancement of Science

The old-established British Association for the Advancement of Science played – as it still does – a significant role in British science, including astronomy. Its annual meetings, held each year in a different place, were open to all those interested in science, including women; foreign scientists were also attracted to its gatherings. The office of President of the British Association, held for one year, was a high honour, and the Presidential address represented an important pronouncement on some current scientific topic. William Huggins was President for 1891 – an event that greatly excited Agnes Clerke when she first heard from Mrs Huggins that he had been invited to take on the position, though he had been reluctant to accept at first because of the difficulty of making an effective summing of astronomy during a 'state of rapid transition and advance'.[10] Agnes Clerke did not normally attend the Association meetings, but on this occasion she and her sister (as well as Mrs Huggins) turned up in Cardiff for the meeting of August 1891.

William Huggins had no greater admirer than Agnes Clerke. On the morning following the Presidential address[11] she wrote off a long report to Gill.

> The address last night [on astronomical spectroscopy] was, as you can imagine, most impressive, if aimed over the heads of most of the audience. . . . it was a noble and memorable discourse giving the upshot of a long life spent in earnest and disinterested investigation; and I hope that it may lead to a further development of thought and enquiry about the wonderful subjects he dealt with.[12]

The young George Ellery Hale

Also at the Cardiff meeting was the 23-year-old American astronomer George Ellery Hale who had reviewed *The System of the Stars* in *Publications of the Astronomical Society of the Pacific* only a few months previously (Figure 8.1). Agnes Clerke was amazed at his youth ('only a boy') – as he was at her years (she was now almost 50). 'He was rather frightened at first', wrote Agnes Clerke to Gill in the same letter, 'not being sure that he was not to "catch it" for his searching but honest and able review of my book; but we soon became good friends.' Far from being displeased by Hale's comments about her book, Agnes Clerke had taken careful note of them, and many years later told him that she still preserved his review article and had profited from it.

Hale, born into a wealthy family in Chicago, had an absorbing interest in science since childhood.[13] While still a student at MIT he set up his own observatory at the family home in Chicago. He came to know, or introduced himself to, America's leading astronomers, who were all greatly impressed by his enthusiasm and his brilliance. His crowning achievement at this stage in his career was the invention of the spectroheliograph, an instrument which was to be hugely successful for observing solar prominences and the upper layers of the sun.

The principle of the spectroheliograph is that the sun's image is scanned by the slit of a spectroscope while another slit, moving at

Figure 8.1 (opposite) George Ellery Hale as a young man. Courtesy of the Archives, California Institute of Technology.

exactly the same pace, isolates a tiny portion of the spectrum, behaving in effect like a narrow-band filter. A photographic plate behind the second slit then records an image of the sun in the light of that particular wave-band. Hale used the deep violet line K, produced by ionised calcium, which is conspicuously present in solar prominences. The photograph or spectroheliogram thus produced showed not only prominences beyond the limb of the sun but prominences and other features projected on the sun's disk. The idea had come to Hale 'out of the blue' while he was a student; he obtained the first successful photograph in May 1891 and immediately announced it in the American magazine *Sidereal Messenger*. News of his achievement had preceded him to Cardiff, where he read a paper on solar prominence spectroscopy and was treated 'like a Grand Duke' – 'all Dr Huggins' doings, for he is all powerful', he wrote home.

Hale also reported home the 'great discussions' he had with Agnes Clerke. The two took to each other, and were to keep in touch for the rest of Agnes Clerke's life. Like everybody else at the Cardiff meeting, she was much impressed by his 'striking photography of prominences',[14] of which she was to give an excellent account in the third edition of her *History* (1893). Hale talked to her about his dream of an international journal of astrophysics that would bring together the best ideas in physics and astronomy from all over the world – something Agnes Clerke always tried to do in her articles and in her books. She was enthusiastic about his idea.[15] Hale's wish was to be abundantly realised in the *Astrophysical Journal*, founded in 1894 with himself as editor. Meanwhile, he co-edited its short-lived predecessor *Astronomy and Astrophysics*, to which Agnes Clerke subscribed from the beginning[16] and supported with contributions in 1892–93.

Other distinguished astronomers with whom both Hale and Agnes Clerke had discussions in Cardiff were Robert Ball, still in Dublin but soon to move to Cambridge, and Ralph Copeland, now Astronomer Royal for Scotland.[17] With Ball she would have talked about his forthcoming book, *The Story of the Sun*, which she was asked to read in proof and for which she is thanked in the Preface.[18]

Immediately after the Cardiff meeting, Hale, having learned that Henri Deslandres at Meudon Observatory was claiming to have produced spectroheliograms like his own, went to Paris to see for himself, accompanied by Arthur Cowper Ranyard, a prominent amateur figure in the Royal Astronomical Society and editor of the popular journal *Knowledge*. The matter of priority for the invention of the spectroheliograph caused friction between Hale and Deslandres, of which Agnes Clerke was not aware for many years. After this visit, Hale and his wife made a quick tour of scientific centres on the Continent, before hastening home. There was no time to visit the Clerkes as they had hoped.

> It was with the greatest regret that we heard of your swift passage through London and disappearance across the Atlantic. We should indeed have liked to renew our short, but pleasant, acquaintance with you and Mrs Hale. Nor do we at all doubt your kind willingness to do so, had it been possible. . . . I fully sympathise with your anxiety to get back to the scene of action at your observatory. There, I feel confident, some of the discoveries of the future will be made, for one may safely predict that your initial successes will prove only preludes to what is to come.[19]

Two years later Agnes Clerke met Hale again, by which time he had risen meteorically and was in charge of the magnificent new Yerkes observatory under construction by the University of Chicago. Throughout 1892 he had been able to produce beautiful photographs of solar prominences. One of these, a striking photograph of an artificially eclipsed sun with its ring of prominences, was shown to Agnes Clerke by Ranyard at the Royal Astronomical Society. She asked and received Hale's permission to reproduce it in the new edition of her *History*, just about to be published. In the same letter she was able to congratulate him 'on having found a millionaire and on the splendid prospects for research which his munificence opens up.'[20] This was Charles Yerkes, who financed and gave his name to what became the great Yerkes Observatory of the University of Chicago.

Agnes Clerke and her family were to see Hale and his wife soon again. They called on them in 1894 on their way to Germany where Hale planned a period of study. 'Hale is splendid!', she told Gill, 'He has the self-denial to leave his cherished instruments and devote a season

to studying physics simultaneously with German in Berlin. His hand-
some young wife who cares only for his career accompanies him,
though she is said to hate foreign languages.'[21] In fact, Hale did not per-
severe with his studies towards a doctorate in Germany, but returned to
his own researches at Chicago. On their way home, the Hales visited
the Clerkes again, this time in Surbiton on the river Thames, then
beyond the city of London, where the family used to take holidays.[22]
Meanwhile, John Brashear, the gifted telescope-maker who had constructed Hale's spectroheliograph, had been in London and had met
Agnes Clerke at a meeting of the British Astronomical Association.[23]

The Royal Astronomical Society and the Royal Society

Though the British Astronomical Association provided a forum for
women astronomers, and though they were all warm supporters of their
new society, some women members were still conscious of their exclusion from the Royal Astronomical Society. The question of admitting
women to the Society had been discussed and rejected in 1886. In 1892 the
matter was brought up again, when three women were proposed for election according to the rules. Agnes Clerke, never an activist, was not
among them. First of the three was Elizabeth Brown; the other two, Alice
Everett and Annie Russell, were 'lady computers' at the Royal
Observatory at Greenwich.[24] They were highly competent young women
who had done well in the Cambridge mathematical tripos examination,
had worked hard at Greenwich and were now in every sense of the word
professional astronomers. Their case was taken up by Downing and
Maunder, senior members of staff at the Royal Observatory, who spon-
sored their nominations. Maunder was the Society's new secretary and
Downing its outgoing secretary. A ballot was taken at the April meeting
of the society which, surprisingly, resulted in their failing to receive the
requisite number of votes. In spite of this hurtful setback, compounded by
a flippant report of the proceedings published in the *Observatory* magazine, the two young women went on to make interesting careers for themselves (Chapter 13). Miss Brown died in 1889, aged 69.

Had the British Astronomical Association not existed, it is possible that the fight for the excluded women would have continued. Regrettably it was to be 23 years before women became eligible to join. Instead, 'a mild plaister was offered to the possibly wounded feelings of the three ladies' in the form of cards of admission to meetings issued by the President 'to such persons as it may be thought desirable to admit.'[25] Agnes Clerke and her sister Ellen qualified, and lost no time in taking advantage of the privilege. They were present at the very next meeting of December 1892, and attended regularly thereafter. Agnes was happy to listen in silence. She described her reaction when, after Hale's lecture in 1893 at which he displayed his remarkable spectro-heliograms, she was asked by the President to say a few words: 'I compelled myself to stand up, but I was so anxious to sit down again that I did not give myself time to explain what I meant to say, so the assembly was not much the wiser for my attempt at public speaking.'[26] However, at meetings of the society she was frequently seen 'surrounded by leading astronomers, genuinely keen to hear her opinion on some knotty point.'[27]

The Royal Society, with all-male fellowship, also made some concessions to women. From as early as 1876 it allowed women guests to be present at certain functions of a social kind, formally at the invitation of the President. Agnes Clerke attended such a function for the first time in 1887 as the guest of Norman Lockyer. In 1895, following a review of its policies on visitors to meetings, the Royal Society decreed that each fellow be allowed to invite one guest, who might be a lady, to meetings.[28] Agnes Clerke never lacked an invitation to lectures and social functions at the Royal Society, and made a point of reporting on matters of astronomical interest at the Society's meetings in the *Observatory* magazine.

The Actonian Prize of the Royal Institution

The Royal Institution, founded in 1799 to introduce useful science to the public, was and remains famous for its Friday evening discourses and the annual Christmas lectures delivered by renowned scientists. Here the Clerke sisters could hear and see many of the experts of the day and keep in touch with developments in various fields. Gill, Huggins and Ball were among astronomers who lectured at the Royal Institution in that period. A constant presence was James Dewar, Jacksonian professor of Natural Philosophy at Cambridge who since 1877 had simultaneously occupied the chair of chemistry at the Royal Institution, where he performed his pioneering researches into the liquefaction of gases. Agnes Clerke and her sister were first introduced to Dewar by Huggins after a Friday lecture on telescopes delivered by A.A. Common in June 1890; with his 'delightful Scotch hospitality' Dewar invited them upstairs for tea where they met 'Lord Rosse and all the swells'.[29]

This was the beginning of a more intimate and valued relationship with the Institution. Three years later Agnes was awarded its Actonian Prize. The endowment in memory of Samuel Acton, an architect, made by his widow in 1838, provided for a prize of a hundred guineas (£105) to be awarded every seven years for the best essay upon 'The Benificence of the Almighty' as illustrated by discoveries in science. Agnes Clerke's obvious reverence for the Creator in her writings certainly made her a deserving candidate. The prize carried considerable prestige. The previous recipient in 1866 was Sir George Gabriel Stokes, the Irish mathematician, for his work on light. Agnes Clerke seems to have been taken by surprise by the award. In accepting she expressed her deep sense of honour to the managers of the Institution: 'That they should have thought my books worthy of so high a recompense must always remain a source of greatest pleasure to me', she wrote.[30] Edward Holden heard the news from Dr Garnett and wrote his congratulations in his usual warm style.[31] She put the prize money to good account by buying herself a Bluthner piano.[32]

Visitors from abroad

Agnes Clerke's widening contacts included foreign astronomers visiting London who experienced the 'genial hospitality'[33] of the Clerke family for lunch or tea at their home in Redcliffe Square. These, as reported in the flow of letters to the Gills, included Gill's collaborators on his international projects – Arthur Auwers of Germany, Jacob

Kapteyn of the Netherlands, Simon Newcomb of the US Naval Observatory and his young countryman William Elkin who had worked for two years at the Cape. Auwers, who came in October 1890, was accompanied by the eminent Vogel.

> Do you know that I have had the honour of making Dr Auwers' acquaintance? He and Dr Vogel, his travelling companion, most kindly came to lunch and gave us some hours of their valuable time. The latter fairly intimidated me by his innate force [of] intellect. My alarm was quite [dispelled?] by the sense of his real kindness and sterling worth . . . He [Auwers] is a man that any other man, however gifted, may be proud to have made his friend. You can imagine how glad we both were to renew our reminiscences of the Cape.[34]

Auwers was Germany's foremost authority on classical positional astronomy. The discussion with Vogel on that occasion centred on the new large telescope for Potsdam. Kapteyn, an astronomer with no observatory of his own, volunteered to measure Gill's catalogue plates and set up a laboratory in Groningen specially for that purpose.

Newcomb paid one of his many visits in 1892 after a trip to Dublin when he was an honoured guest at the Tercentenary of Trinity College, representing Johns Hopkins University.[35] They had a long discussion about various astronomical matters. 'I promised to tell you how well and robust he looked', Agnes Clerke reported to Gill. 'He carries a crutch but, as we told him, it must be for ornament since he appears to walk quite well without it, and he looks strong and young for the years that I suppose must be his [he was in fact only 59]. But Americans begin so young that they have often done a good average life's work while Europeans are looking about them waiting to start.'[36] Herbert Newton of Yale, well-known for his work on correlating meteor showers with past comets, was another of her American guests. 'I would scarcely allow anyone to interfere with my monopoly of the Professor who was fascinating on the subject of meteorites.'[37] So was Thomas Jefferson J. See, taken by Cowper Ranyard to visit the Clerke family at their home in 1892 after the British Association meeting.[38]

From Germany, an important visitor and friend was Max Wolf of

Heidelberg who, like Edward Holden, appears to have introduced himself to Agnes Clerke by sending her copies of his first striking celestial photographs. In 1891, at the age of 28, Wolf, a mathematician at Heidelberg University and an ardent amateur astronomer since boyhood, emulating the technique pioneered by Barnard at Lick, began to photograph Milky Way fields using a wide-angle lens. His first photographs were of fields in Cygnus, showing bright nebulosities and dark spaces, believed at that time to represent areas devoid of stars. He was the discoverer of the star cloud known from its shape as the North America nebula, and pioneered the method of discovering minor planets (the first found in December 1891) on long-exposure photographs of ecliptic fields, where they showed up as little streaks. This work was done at his private observatory; later, in 1893, having made a name for himself in astronomy, he was made Director of the newly built Heidelberg Observatory on Königstuhl.

Agnes Clerke, responding to Wolf's gift in June 1892, wrote: 'Pray accept my grateful thanks for the unexpected gift of five of your splendid photographs. They are triumphs of the art of short-focus pictures, and few can appreciate their profound importance as indexes to the ultimate structure of the Milky Way. I consider myself indeed fortunate to possess, through your kindness, such fine specimens of them.'[39] She could not 'resist the temptation' of asking if he would consider photographing the double cluster in Perseus, notable for its red stars, 'with a view to developing possible nebulous relationships of the stars.' (There is no indication that Wolf carried out her request: the photograph of the double cluster in the second edition of *System of the Stars* is by Barnard.) It was the beginning of a correspondence, with many queries on her side, that went on throughout her life.[40] In October 1893 Wolf met the Clerke family when he visited them on his way home from America. He had been to Lick, and, as Ellen relayed to Holden, reported 'the great telescope a model of perfection'.[41] The first of Agnes Clerke's letters to Wolf are in English, but many others are in German, a language in which she was fluent. Though mainly concerned with astronomy, we find among them Christmas and New Year greetings, and a note of congratulation (in German) to Wolf and his wife on the birth of their son in 1898.

Ellen

Agnes Clerke was usually accompanied at scientific gatherings by her sister Ellen (Figure 8.2). As well as being a writer on European literature and foreign affairs Ellen was a competent scientific journalist, special-ising in geography and anthropology. She was a member of the Manchester Geographical Society, founded in 1885 by a group that included Vaughan, then Bishop of Salford, a great enthusiast, who was a Vice-President. The Society, the second only in the country after the Royal Geographical Society in London, aimed, with noted success, at spreading knowledge of geography for commercial and educational purposes at home, and of culture, including the work of Christian mis-sions abroad.[42] It attracted large audiences to its meetings to hear lec-tures from academics, explorers and missionaries. It prided itself on welcoming women as members from the outset – unlike the London Society – and in inviting women lecturers. Ellen contributed to the society's journal, but, according to the records, never addressed the meeting in person. She did, however, read a short paper at the Geographical Section of the British Association meeting at Cardiff in 1891 on the subject of aboriginal Australians, drawn from accounts of Benedictine Missionary priests and presented as 'a most striking refuta-tion of the generally received belief' about them.[43] This talk, as pointed out by Mary Creese, 'is perhaps as interesting now for the light it sheds on current attitudes towards non-Europeans as for its information on aboriginal life and culture'.[44] The subject is typical of the material regu-larly published by her at greater depth and length in the Dublin Review.

On wider issues, Ellen was the author of a major article on the Dock labourers strike of 1889 which was resolved by the intervention of Cardinal Manning.[45] The material nominally under review were two official Government Reports, and the first volume – just published – of Labour and Life of the People, dealing with conditions in London's East End, by Charles Booth,[46] a radical writer on social problems. In this excellent essay Ellen described with great feeling current economic policies and theories, the wars of class against class and people against people, and 'the uncompromising and unedifying motto [of] Vae Vietis! or the weakest to the wall'. 'The struggle for existence', she wrote,

Figure 8.2 Ellen Clerke. Royal Astronomical Society.

'which eliminates the weak and secures the survival of the fittest, may be regarded with equanimity in the brute and vegetable worlds, but when enacted by the surging human mob at the dock gates of East London, it immediately brings home to us the conviction that there is something wrong in the scheme of social existence that produces it'. How to deal with it, was the question. She described the economic problem of foreign competition, the trap in which low paid workers fell, including wives compelled to work to supplement their husbands' incomes, and the noble efforts of charitable organisations. Like Agnes in her own field, Ellen was a stickler for facts; here she gives statistics of employment at various levels in the docks (there were 50,000 men involved) and other details. The essay ends with praise for Cardinal Manning's 'kindly human sympathy for all, but more especially for the poor of the great city'.

Also in 1889, the year of its centenary, an article by Ellen entitled 'The principles of '89' was a strong attack on the French Revolution and all it stood for.[47] One of the works under review was Alexis de Tocqueville's book on pre-Revolution France translated by Henry Reeve and two in French on contemporary France. Ellen's article begins by criticising 'the fashion with many modern historians to treat the revolution of 1789 as a beneficent movement for reform'. She would have none of this. 'It was from its initiation an organised attack, captained and led by sworn conspirators, on all pre-existing institutions, beginning with religion as the cornerstone of society.'

Ellen Clerke was not a mere freelance writer. About 1885 – the year in which Agnes published her *History* – she joined the permanent staff of the influential Catholic weekly newspaper, the *Tablet*. The *Tablet* had a link with the *Dublin Review* through the Westminster Catholic hierarchy and was known for its generally conservative stance on matters religious and political. Details of her contributions to the pages of the *Tablet* (which were unsigned) cannot now be traced (the archives of the *Tablet* were lost in the bombing of London in the Second World War),[48] but she was highly regarded as a commentator on Italian and German politics.[49] She served on the permanent staff, with occasional editorial duties, under John Snead-Cox who became editor in 1884, and continued until her death.[50] Snead-Cox, a personal friend of the Clerkes (mentioned in a letter of Agnes to Edward Holden[51]) was a first cousin of Vaughan, prime mover of the *Dublin Review* who was soon to be Cardinal. Snead-Cox, who was to edit the journal for 36 years, is credited with having raised the *Tablet* to the rank of a first-class periodical.[52] The literary side was strengthened, and scholars recruited to contribute articles of lasting value, Ellen being no doubt among them. Some women readers criticised Snead-Cox for his 'stiff opposition to the Women's Suffrage Movement'.[53] The *Dublin Review* was also lukewarm on women's issues, but it was not a matter to trouble either of the Clerkes.

As to writings on astronomy, Ellen contributed occasionally to the *Observatory* magazine and to the *Journal of the British Astronomical Association*. An unsigned article on the star Algol,[54] describing its history and the derivation of its Arab name (meaning the Demon) is quoted by Agnes over her sister's name in the later editions of her *History*. In 1892 she published a 40-page 'popular brochure' on Jupiter[55] (whose fifth satellite had just been discovered at Lick by Barnard). It merited a brief word of recommendation in *Knowledge* and was praised in *Nature* for being 'popular yet accurate' while the ever-helpful Edward Holden gave it a profuse notice in *Publications of the Astronomical Society of the Pacific*, calling it 'a capital little pamphlet, equally removed from dullness and from the faintest trace of "smart writing"'.[56] A copy also went to Hale, whom Ellen knew well through her sister.[57] The book, which sold 700 copies in the first three months,[58] was followed by another, in the same vein, on Venus.[59] Ellen was disappointed that the publishers 'were not encouraged to undertake a further venture'.[60]

Agnes' 50th birthday

Agnes Clerke reached her 50th birthday in February 1892. Once dubbed the Unknown, she was now a successful writer, a respected commentator on astronomy, and the object of many warm friendships. The hospitable Clerke home, like the Hugginses' Tulse Hill Observatory, seems to have become part of the visiting astronomers' trail in London.

9 Homer, the Herschels and a revised *History*

Studies in Homer

Another revision of her *History of Astronomy* was now due, but before tackling it Agnes Clerke turned 'as a sort of recreation'[1] to preparing a little volume on Homer[2] which came out early in 1892. 'Greek enough to read the *Iliad* and the *Odyssey* in the original can be learned with comparative ease', she wrote, 'and what trouble there may be in its acquisition meets an ample reward in mental profit and enjoyment of a high order.' As an adult she had made herself proficient in Greek with a volume of Homer and a dictionary – the best way, she thought, of mastering a dead language, much better than spending months on the disheartening drudgery of committing grammar to memory (though keeping a Grammar always by for reference).[3] Her intention in writing about Homer was to share her own pleasure with her readers and to promote a 'non-erudite' study of Homer's 'noble poetic monuments' at a time when archaeological discoveries were uncovering actual physical evidence for the existence of ancient civilisations.[4]

Familiar Studies in Homer was a collection of essays on aspects of the Hellenic world as gleaned from the *Iliad* and the *Odyssey*. It included one on Homeric Astronomy based on an article originally published in *Nature*. Other delightful essays dealt with such matters as Homer's dogs and horses, trees and flowers, herbs and food. Agnes Clerke's discussion of astronomical allusions in Homer's poems is a scholarly work that anticipates recent research on the same topic. Homeric ideas regarding the heavenly bodies, she explained, were of the simplest description. The poet's profession was not science but song. She pointed to the interesting fact that abnormal astronomical events are scarcely noticed in the Homeric poems; comets, she wrote, 'have left not the suspicion of a trace in these early songs'. In her quotations from Homer Agnes Clerke chose from a variety of versions in English including some passages translated by herself. (During the latest appearance of Halley's comet (1986) various commentators, relying, like their predecessors in 1910, on a free rendering of certain verses in the *Iliad* by the poet Alexander Pope in the early eighteenth century, mistakenly claimed that the comet had been on record since Homeric times[5]. Agnes Clerke being familiar with the original Greek made no such mistake.)

Familiar Studies in Homer brought Agnes Clerke a communication from Samuel Butler, best known as the author of *Erewhon*, the story of an imaginary country shut off from the rest of the world. Among the fields of interest of Butler's enquiring mind were the classics. He studied the problem of the authorship of the *Iliad* and the *Odyssey* and came to the conclusion that the two poems were written by different people, the author of the *Odyssey* being, he claimed from a close perusal of the text, a woman from Sicily. This view was first put forward in a magazine article in 1892 and eventually in a book *The Authoress of the Odyssey* (1897). Butler, a habitué of the British Museum sent Agnes Clerke (at Richard Garnett's suggestion) a copy of his own pamphlet *The Humour of Homer* (1892). Her charming response, though not Butler's letter, is preserved:[6]

> You are very kind in adopting Dr Garnett's suggestion and I beg to offer you my best thanks for your Lecture on the Humour of Homer. I had read your communication in the *Athenaeum* on the Geography of the Odyssey and your strikingly original view as to the authorship of that Poem. I, as a woman, ought to feel grateful, since a greater compliment to the female intellect could scarcely be paid.

Butler, who spent ten years of his life developing his theory, was not taken seriously by the classical scholars of the day though some later literary men including George Bernard Shaw were impressed.[7]

The third edition of the *History*

Since completing *System of the Stars*, Agnes Clerke had been preparing a revised third edition of her *History* and carrying on her serious correspondence with Holden and Gill, in which the latest astronomical events were mulled over. She asked Holden quite particularly for criticisms and comments. He had made notes in his copy of the first edition as he 'rode between Berkeley and San Francisco', and Agnes Clerke asked if she could see these, to help with her revision. He, however, felt that to send them 'in cold blood all across the ocean would give them an importance which they do not deserve. So, if you please, I shall wait till I do have the pleasure of seeing you in London (or here) and of presenting my suggestions for the nth edition!'.[8] In fact, the two were never to meet.

The interval of six years since the second edition had been a time of 'marvellously fruitful astronomical discoveries'; the new edition was 'executed without stint of care or pains' and offered in the 'hope that the remodelled work will enjoy no less favour than was shown to its earlier issues'.[9]

Naturally, the writing of *System of the Stars*, published in 1890, had provided much new material. A major development since then was that rare event, the discovery of a nova (new star), Nova Aurigae, in the constellation of the same name, announced, as it happened, on her fiftieth birthday. The nova was discovered by an amateur astronomer in Edinburgh on 1 February 1892 and made known to Ralph Copeland at the Royal Observatory Edinburgh who circulated the news among the astronomical community by telegram. It transpired that the star had in fact been recorded photographically in December at Harvard at a spot where, on a photograph taken 24 hours earlier at Heidelberg by Max Wolf, nothing had shown up. Other Harvard photographs showed that it had reached maximum brightness (4.2 magnitude) on December 20. Only one other nova had been recorded since the famous nova of 1866 (T Coronae), whose spectrum was observed visually by William Huggins. Nova Aurigae (now known as T Aurigae) was the first bright nova to be well observed spectroscopically by photography, several observatories being by now equipped for such observations. The nova generated enormous excitement and activity. The spectrum turned out to be double: there was a normal dark-line spectrum, and a separate emission-line one. More extraordinarily, the two spectra showed vastly different Doppler motions, one component receding and the other approaching, at very high speeds. At one stage, the relative motion of the two spectra was observed to be 500 miles per second – a bafflingly high figure, quite beyond what could be explained by the usual model of a double star in which the stars orbit each other. Agnes Clerke wrote a review of the observations (read out on her behalf to the Toronto Astronomical Society and later published),[10] confidently asserting that the only possible explanation was a close encounter between two stars already moving through space at high relative speeds – a 'tidal' theory set out by William Huggins in a lecture to the Royal Institution. She had been too hasty in adopting Huggins' explanation. In August the object transformed itself into a faint star within a nebula, and Agnes Clerke was forced to admit that 'the collision theory collapsed under the weight of facts'.[11] The true explanation of a nova, not reached at that time, is the throwing off of an outer shell, the bright-line element in the spectrum arising from the shell which shows a motion of approach relative to the star itself. No second star is involved.

The book was illustrated by some very recent photographs, including Hale's spectroheliographs of solar prominences taken at his Kenwood Observatory in Chicago, Barnard's famous photograph of the Milky Way in Sagittarius (Figure 7.2) and the Hugginses' beautiful spectrum of Nova Aurigae (Figure 9.1). There were useful and interesting appendices (Agnes Clerke was an indefatigable recorder of data) and an extended chronological table of some 400 astronomical milestones from 1774 to 1893, including the deaths of important astronomers.

Reviews

The (unsigned) review in the *Observatory*[12] gave an assessment which, even a century later, seems well deserved.

> She is to be congratulated on the happy tact displayed in selecting the additions necessary to make it complete to date. All that the reviewers have said in favour of previous editions we can now endorse, and can heartily recommend to the ordinary reader as well as to the astronomer, as a book which will give a clear grasp of the progress and present state of astronomical science.

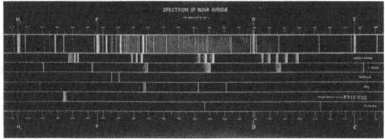

Figure 9.1 Spectrum of Nova Aurigae by William and Margaret Huggins. Top: negative photograph of the blue end (lower spectrum) compared with the spectrum of Sirius. Bottom: drawing of the visually observed red part of the spectrum (middle spectrum, showing two emission lines) compared with the spectra of various laboratory sources.

The American *Astronomy and Astrophysics,* edited by William W. Payne and George Ellery Hale, said 'The book reads like a romance, yet it deals with the present century's contributions to astronomy in no sensational way'. The review pointed out the recent researches covered – the eclipse of 1889, theories of the corona, the fifth satellite of Jupiter, Nova Aurigae 1892 – which brought the book thoroughly up to date. 'To every student of astronomy and especially to the rapidly increasing number whose chief interest is in the progress of astrophysical investigation, we heartily recommend this book.'[13]

Clash with *Nature*

The review in *Nature,*[14] however, was less friendly. Certain alleged but in fact not genuine errors were picked out and 'refuted'. Praise for the book as a whole was nullified by the remark that 'it can hardly be said that the strict impartiality which should characterise a history of astronomy has been exercised when an event of such local interest as *Lecture by Dr Huggins on Nova Aurigae at the Royal Institution 1893* is recorded, while the announcement of the duplex nature of the lines [by Vogel] in the spectrum of the same nova is unmentioned in the [chronological] table'. (Vogel's radial velocities were superior to Huggins'. Agnes Clerke subsequently dropped the Huggins item from her chronology.) 'The merits of the volume are now so well known', the review went on, 'that it is quite unnecessary to expatiate upon them. It seems to us, however, that if Miss Clerke were more of a historian and less of a partisan her work would be of higher value'. The attack was really aimed at Huggins, who disagreed with Lockyer on the question of the meteoritic theory of the origin of the stars, though the latter's name was not mentioned. The anonymous reviewer was undoubtedly Richard Gregory, one of Lockyer's assistants who, like Fowler, had spent some years working in Lockyer's spectroscopic laboratory. Gregory had begun writing for *Nature* in 1890, and had recently (1893) become assistant editor and regular book reviewer.

It is hardly a coincidence that Agnes Clerke's contributions to *Nature* ceased at that point. This was surely a loss to *Nature*, but a greater loss to her. *Nature*, celebrating its twenty-fifth anniversary in 1895, was now 'the most influential journal of science in Europe'.[15] Agnes was always ready to accept criticism. One wonders if pressure was exercised by the spirited Margaret Huggins on her friend in the affair. Agnes began instead to contribute to the popular and attractive journal *Knowledge* now in the hands of her good friend Cowper Ranyard who had taken on the task after the death of R.A. Proctor in 1888. Ranyard, a lawyer by profession and a keen amateur astronomer, was for most of his life a member of Council of the Royal Astronomical Society. He died in 1894.[16] 'We feel very sad about Mr Ranyard', wrote Agnes Clerke to the Gills shortly before his death. 'The doctors do not know what is the matter with him. He has put a great deal of money into *Knowledge*, which is now beginning to pay its expenses, but I fear very few more numbers will be published.'[17] In fact, the journal was taken over by Baden Baden-Powell, a retired army officer of the famous family of that name, with E.W. Maunder acting as editor of the astronomical sections.

The one-system universe reaffirmed

Agnes' personal friendship with Cowper Ranyard did not deter her from penning a sharply critical account of his cosmological views in an anonymous essay in the *Edinburgh Review* in 1893.[18] The principal work under discussion was *Proctor's Old and New Astronomy*, the book which Ranyard had completed after the author's death. Proctor, a talented scientist and a prolific popular writer, died in tragic circumstances in New York in 1888 at the age of only 51. *Old and New Astronomy*, which he had been working on spasmodically for a quarter of a century, was to be his magnum opus, 'a work which should embody all the best results of his life's scientific activity'. Beginning in 1887, it had been planned to appear in twelve instalments of which seven had been published before his death. The remaining sections were supplied by Ranyard, and the entire work published in 1892.[19] It was Proctor's fifty-seventh book.

Though Agnes admitted that Proctor was a controversialist (in her early days she had witnessed his onslaught on Edward Holden) who 'seemed equally impressed with the truth of his own convictions and with a sense of the amazing dullness of his opponents', she nevertheless was a great admirer of his mathematical talent and of his industry. He had been educated at Cambridge, but had unexpectedly achieved only a poor degree – a circumstance attributed by Agnes to 'the development of a taste for athletics and his marriage, while still an undergraduate, to an Irish lady whom he met during one of his vacation trips to the Continent'. Driven by circumstances to earn his living by popular writing and lecturing, he acquired a huge following as a prolific author and speaker on both sides of the Atlantic. But in addition to being 'unrivalled as an interpreter to the unlearned of the geometrical facts of astronomy' he was an accomplished cartographer and statistician.

Proctor's work as an astronomical cartographer involved two enormously laborious projects concerned with the distribution of stars and nebulae in the sky. One was a plot on equal surface projection of all 342,000 stars in Argelander's great catalogue (the *Bonner Durchmusterung*).[20] The other was a similar plot of all the 'irresolvable' nebulae listed by John Herschel in both hemispheres, 4,000 in all.[21] These showed two distinctly different distributions from which Proctor concluded back in 1869 that the stars and the nebulae were part of the same scheme: had the nebulae been distributed uniformly throughout space their distribution would not have shown such preference. It was to Agnes Clerke, as she often repeated, the most convincing evidence of a one-system universe.

Ranyard, however, believed differently. Contrary to what Proctor would presumably have written, he declared in favour of an infinite universe and multiple other inhabited worlds. He countered the argument that the sky was not bright all over by the hypothesis of light absorption in space, and suggested that the dark lanes in the Milky Way were due to actual obscuration. Agnes Clerke lashed out in her anonymous review in a manner that would have been impossible face to face. 'In order to nullify the effect of this *reductio ad absurdum* of an infinite universe', she wrote, 'appeal was made to a conjectural extinction of light in space; and Mr Ranyard trusts further, for the protection from the intolerable glare of unnumbered suns, to the intercepting effects of dark masses, which, for anything we know to the contrary, may abound in celestial tracts equally with bright ones. But there is not the slightest sign that the apparent distribution of the stars is essentially modified by either of these causes.' 'There can be no question but that the stars and nebulae together form a definite structure; and a definite structure can only be composed of a finite number of objects. This we take to be self-evident.' As to the dark gaps in the Milky Way – they are 'undoubtedly what they seem; they are obscure simply because they are destitute of stars. They are negative entities. Mr Ranyard, however, takes a different view. Openings, clefts and chasms ... represent opaque formations projected upon and cutting out the light from starry strata lying behind them. We are not, it is true, in a position to deny the interpositions of this kind. Dark stars certainly exist. Why not dark nebulae? For the purpose in hand, nevertheless, they appear to us superfluous, and even embarrassing.' Ranyard's inclination was right about the obscuring matter in the plane of the Milky Way, but wrong about obscuration in an infinite universe (though he was in good company, since the idea had first come from Wilhelm Struve of Pulkovo as far back as 1847). Time would show that Agnes Clerke herself was right on neither count.

Visit to Daramona

The Clerke sisters are likely to have paid visits to family and friends in Ireland from time to time from London. One visit of scientific interest and importance took place in April 1893, when they spent a week at Daramona, the home and observatory of the astronomer William Edward Wilson, in Co. Westmeath.[22] It was no doubt a well-appreciated breathing-space for Agnes, while the manuscript of *A History* was in press. Wilson, the son of an affluent and intellectual landowner, had been educated at home, and as a young man had developed a keen interest in astronomy. He began with a 12-inch telescope, but in 1881 bought from Grubb, the Dublin opticians, an excellent 24-inch silvered-glass mirror instrument which he had re-mounted in 1892.[23] From that date he carried out a succession of fruitful observations, some in collaboration with other scientists, which placed his observatory among the notable amateur establishments of the nineteenth century. A principal interest was the effective temperature of the sun, measured by comparing the solar radiation with that of an electrically heated strip of metal, the first realistic measurement of that important datum. His final result, of about 6,800 K, was remarkably close to the true modern value. 'I think the result will turn out the most reliable yet obtained', wrote Agnes Clerke to Gill following this visit. 'He used a duplex radio micrometer and adopted, after careful testing, Stefan's Law of radiation. If the photosphere be really of this heat it cannot be made of solid carbon particles since according to the latest research, carbon volatiles at 4,000°.'[24] A collaborator in Wilson's work was Arthur Rambaut, successor as Director of Dunsink Observatory of Sir Robert Ball from 1892 and later (1897) Savilian Professor of Astronomy at Oxford.

A particular point of interest to astrophysicists in the 1890s was the decrease in the brightness of the sun's surface between the centre and the edge, the so-called limb-darkening, interpreted as being due to an absorbing layer in the sun's atmosphere, what Agnes Clerke called the 'dusky veil'. In 1893–94 Wilson and Rambaut looked for similar brightness variations in sunspots between the centre and the limb.

Wilson also began celestial photography with his 24-inch

reflector in 1893, producing some remarkably fine photographs of nebulae. He and Mrs Wilson visited the Clerkes at their London home in April of that year, and in June the Clerkes were the Wilsons' guests in Daramona, where they were joined by Rambaut and his wife. Wilson's diary entry for a day during that week – 'Picnic on the lake. Observing at the observatory with Rambaut' – seems to imply that the ladies were left on the lake while the men went to work. But Agnes Clerke would have seen the solar apparatus even if she did not actively take part in the observations, and may well have seen the beginning of Wilson's celestial photography.

Wilson was related to the distinguished Irish literary family, the Edgeworths, and before leaving Daramona the Clerkes paid a visit to Edgeworthstown House, just a few miles away, where in earlier days the celebrated writer Maria Edgeworth[25] welcomed literary luminaries including Sir Walter Scott and William Wordsworth and scientists such as Sir John Herschel and Sir David Brewster. The Wilsons' nephew, Kenneth Edgeworth, then a boy of 12, who acquired a serious interest in astronomy from his uncle, has become recognised in recent times for his theoretical work on the so-called Edgeworth–Kuiper belt of asteroids.[26]

Agnes Clerke was to devote an entire chapter to the subject of the solar temperature in her later book *Problems in Astrophysics*, giving particular attention to Wilson and Rambaut's work. Indeed, she was the only contemporary commentator to do this.[27] In 1900 Wilson published a book on the astronomical and physical researches made at Daramona[28] which was privately printed with very limited circulation.[29] Agnes Clerke was one of the privileged recipients of a copy of this beautiful and rare volume, a gift which she 'prized exceedingly, both for personal and scientific reasons'. As to the photographs, they were, she said 'simply exquisite, and printed to a perfection rarely achieved'.[30] She would use some of them to illustrate her books, the only writer besides Wilson himself to do so.

The Wilsons visited the Clerkes at their home in Redcliffe Square on subsequent occasions, including one, in 1900, when they had the experience of meeting David Gill – 'an interesting man and a great talker'.

The Herschels and Modern Astronomy

Agnes Clerke's next book, published two years after the revised *History*, was of a different genre. *The Herschels and Modern Astronomy*[31] was an account of the lives and work of William, Caroline and John Herschel, one in a series of popular scientific biographies under the general editorship of the spectroscopist Sir Henry Roscoe. This piece of work was undoubtedly a pleasure. Agnes Clerke owed her early enthusiasm for modern astronomy to John Herschel's *Outlines of Astronomy* which she read as a child with her father, and had taken the researches of William Herschel on the structure of the universe as the starting point of her *History*.

Her study of the Herschels had been long planned. As early as 1889, shortly after her return from the Cape, Agnes Clerke visited the Herschel family home at Collingwood, where John Herschel's son Sir William and daughter Isabella allowed her to examine the Herschel papers.[32] It was serious work: Agnes Clerke was to be responsible for the excellent Herschel entries in the *Dictionary of National Biography*.

The *Herschels*, a thoroughly delightful book, recounted for the general reader the fascinating personal lives of that extraordinary family, while at the same time explaining the significant place of the elder Herschel in the history of astronomy. One entire chapter on Caroline, the 'Cinderella sister', earned particular praise, being probably the first specific account of her as a person in her own right. 'If this book has done nothing else', wrote the reviewer in the *Observatory*, 'it certainly has shown in a lovable light the character of Miss Caroline Herschel.'[33]

Agnes Clerke sent a copy of the book to her old friend Henry Reeve, who was still editing the *Edinburgh Review* in spite of his great age.

> Many thanks, my dear Miss Clerke, for your elegant and instructive Life of the Herschels; they could not have had a more accomplished biographer, if they had waited for it another century. . . . Tomorrow is my 82nd birthday – probably the last. But I am not ill, only feeble and tired of living so long.[34]

The letter was one of the last written with his own hand; he died the following year.

Invitation to join an eclipse expedition

A total eclipse of the sun visible from the Arctic regions occurred on 9 August 1896 and attracted many expeditions from all over the world. The British Astronomical Association organised an expedition to a point on the east coast of Finnmark, the first such venture by the society. This party of fifty-eight amateurs, including several ladies, led by Maunder, joined the official British Government expedition, making up a total of 165 persons who travelled together by ship and land.[35] A small group, which was to include Agnes Clerke and David Gill, neither of whom in the event took part, was invited to travel in the *Otaria*, the private yacht of Sir George Baden-Powell, a wealthy Member of Parliament and noted yachtsman.

Baden-Powell, brother of the future Lord Baden-Powell, hero of the Boer War campaign and founder of the Boy Scout movement (another brother was the new editor of *Knowledge*), had a personal connection with astronomy: his mother was the daughter of the renowned amateur astronomer W.H. Smyth and sister of Charles Piazzi Smyth. Agnes Clerke declined the invitation, pleading pressure of work. She may also have been worried about her mother's failing health, but the chief reason was probably her retiring nature, since, only one week later, she earnestly suggested to her friend Margaret Huggins that the two of them should form a little expedition of their own. 'But it could not be', recorded Mrs Huggins, who perhaps would have found it difficult to leave her husband, now well advanced in years.[36]

It was a missed opportunity for Agnes Clerke to witness the incomparable spectacle for herself. Unlike the main expeditions at Vadsö, which were frustrated by the weather, Baden-Powell's observers at Novaya Zembla had a good view of the eclipse. After the eclipse, Baden-Powell met the famous Norwegian explorer Fridtjof Nansen, just returned from his epic three-year Arctic voyage, and conveyed him in the *Otaria* as far as Christiania.[37] Agnes Clerke made the Arctic expedition the subject of one of her next essays, 'Nansen and the Pole', in the *Edinburgh Review*.[38]

The Clerke family could hear all about the expedition and celebrate its success on the occasion of a visit to London of E.E. Barnard of Lick to receive his award of the gold medal of the Royal Astronomical Society. Agnes Clerke gave a party in his honour, when her guests included Sir George and Lady Baden-Powell and his eclipse companions, the astronomer and physicist Alexander S. Herschel (Sir John's son) and William Shackleton,[39] a member of Lockyer's Solar Physics Observatory. Mr and Mrs Maunder, of the disappointed BAA team, were also present. Agnes Clerke afterwards tried to interest Gill in Shackleton, who evidently wished to follow a permanent career in astronomy:

> I have had some talks with him and think highly of his working
> faculties. His success at the eclipse was due to no mere good luck but
> to sheer skill and intense watchfulness. He has keen intelligence as
> well, and is deeply in earnest . . . I do not say that he is likely to turn out
> anything great, but he ought to be very useful if he gets the chance.[40]

Shackleton's achievement[41] was to obtain the first really successful photograph of the 'flash spectrum', so-called for its brief duration, a phenomenon first observed visually by the American Charles A. Young at the 1870 eclipse in Spain. Shackleton used a slitless spectrograph – one 'destitute of both slit and collimating lens', to use Agnes Clerke's pithy phrase – which showed up the spectrum of the chromosphere as a series of curved images. The word chromosphere was coined by Lockyer after the 1868 eclipse to denote the thin layer above the photosphere which gives rise to the flash spectrum. The idea of dispensing with a slit was Lockyer's, tried out visually by him and the Italian astronomer L. Resphigi at the 1871 eclipse in India, and photographically by Lockyer's team in Africa at the 1893 eclipse. Lockyer – who had been frustrated at Vadsö – was so excited with Shackleton's result that he wrote a letter to the *Times* newspaper, praising the public spirit of Sir George Baden-Powell.[42]

Photography of the flash spectrum by the slitless method became an important element in later eclipse programmes, beginning with the Indian eclipse of 1898 and continuing right up to the space age.

Young gave an account of Shackleton's achievement in the fourth edition (1897) of his famous book *The Sun*,[43] as did Agnes Clerke in her *Problems in Astrophysics*. However, presumably for want of space, it is not mentioned in the last edition of her *History*, and Shackleton's name seems to have been forgotten in later accounts of the subject. Gill did not respond to Agnes Clerke's plea for young Shackleton, who in later years contributed monthly notes on astronomy to *Knowledge*.

The Clerke dinner was part of Barnard's 'brilliant visit' when he was also lionised at Oxford, Cambridge and Greenwich.[44] Agnes was delighted with him, both as a person and as an astronomer. 'Professor Barnard is a simple-hearted straightforward man with something of nobleness under his Nashville exterior.' The award of the gold medal was for his discovery of the first non-Galilean satellite of Jupiter in 1892, but on this visit to England it was his series of photographs of the Milky Way which caused the real sensation. Agnes Clerke saw these when Barnard addressed a specially arranged meeting of the Royal Astronomical Society;[45] to her mind 'they raised some profoundly interesting questions such as whether the nebulously-involved brilliant stars and star-dust of the Milky Way do really coexist in the same region of space'.

Queen Victoria's Diamond Jubilee

The year of Queen Victoria's Diamond Jubilee, 1897, was when a number of illustrious citizens were honoured as part of the celebrations. William Huggins was made a Knight Commander of the Order of the Bath, joining David Gill who had been knighted the previous year. Under the general heading *The Science of the Queen's Reign*, the magazine *Knowledge* published an article by Agnes Clerke entitled 'Sixty years of astronomical research'.[46] In this global overview Britain's home contributions figured only dimly compared with those of Lick, Potsdam and the Cape. The Royal Observatory Greenwich received a perfunctory mention in the shape of a reference to Sir George Airy's time in office (he retired in 1888). Current activities at Greenwich and elsewhere in Britain merited not a single word. The article ended with a salute to 'the sister establishment [of Greenwich] in the southern hemisphere, by the activity of which England's universal dominion over the seas is extended to the skies', and where soon, through the generosity of Mr McClean [recent donor of the Cape's Victoria telescope] 'Imperialism will triumph in the New Astronomy'. The 26-inch Thompson telescope by Grubb erected in Greenwich in 1897 – an instrument of similar power to McClean's – was ignored. No reference was made either to the 24-inch astrograph at Greenwich or to its contribution to the *Carte du Ciel* programme which progressed with exceptional efficiency. Incongruously, the article was illustrated by a lithograph of the ostentatious new Vienna Observatory, with the caption: 'An Ideal Establishment such as we might have, but have not, in England'. Edinburgh's imposing new Royal Observatory building, an excellent example of a modern purpose-built observatory, which had been officially opened less than a year previously (7 April 1896), was nowhere shown or mentioned.[47]

Articles in celebration of the Jubilee were presumably intended to be optimistic if not positively flattering. But in Agnes Clerke's account of the progress of astronomy in the previous 60 years, the Cape alone in all the Empire came in for praise as that period came to a close. The writer was all too obviously banging the drum for David Gill and heavily influenced by his antagonism towards the incumbent Astronomer Royal Sir William Christie. She may also have been indirectly influenced by her friends the Maunders. Walter Maunder, a hardworking member of the Greenwich staff for 25 years, has come to be regarded from his own testimony as an admirer of the Astronomer Royal under whom he was then serving, contrasting his administration exceedingly favourably with that of his predecessor, Sir George Airy.[48] Privately, however, his wife Annie believed that her husband received less than satisfactory treatment at the Royal Observatory.[49]

The verdict of history is that Christie 'had more sense of direction in astronomical trends' than his famous predecessor Airy, and 'presided over the greatest enlargement of the Observatory in staff, in territory, in buildings, and in total light-grasp of its telescopes during the whole of its time at Greenwich'.[50] But there were no reservations in the response of the reviewer in *Observatory* (perhaps Turner, the magazine's editor) to Agnes Clerke's article. 'Miss Clerke has written one of her admirable historical essays on the progress of Astronomy during the reign of her Majesty'.[51] The article was translated into French and published in the popular magazine *Ciel et Terre*.[52]

Death of mother

The year ended sadly for the Clerkes. Their mother died on 17 December, aged 79. She was undoubtedly a woman of strong personality without whom her children, though all in their fifties, felt orphaned. Two years later Agnes still could write of her 'dear mother's death leaving us three to be everything to one another'.[53]

10 The opinion moulder

·

Continued dealings with the Cape

David Gill, tremendous enthusiast that he was, never ceased to complain that his Observatory at the Cape was inadequately funded. The Cape Observatory, like the Royal Observatory at Greenwich, operated under the British Admiralty. In practice this meant that Gill was not independent of the Astronomer Royal in Greenwich: to quote Brian Warner, historian of the Cape Observatory, 'from 1887 [when financial support for his Cape Catalogue was halted] until his retirement Gill had continually to fight the effects of Christie's hostility to almost every proposal that emanated from the Cape.'[1] He kept the catalogue going from that date onwards by contributing half of his own salary to it.

In 1892 Gill declined to be considered for the Chair at Cambridge, which went to his friend Sir Robert Ball from Dublin, insisting that he could do more for astronomy by staying at the Cape. Only a year later, however, in 1893, he was despondent, and Agnes Clerke took upon herself to alert him to a possible opening at home, following the death of Charles Pritchard, Savilian Professor at Oxford. 'You will be surprised at my telegraphing you about the Savilian Professorship' she wrote in a follow-up letter, 'but it seemed to me so important under the circumstances that you knew the appointment is still open that I risked incurring the blame of officiousness. For in the absence of a second assistant you might, I thought, consider your present difficult service to have become impossible, and might prefer seeking elsewhere a field for the high powers which would be entirely wasted on routine work'.[2] She was not sorry, however, when he did not act in the matter: 'I am glad you were not driven to apply for the Oxford post. For you know how enthusiastic I am about your work in the southern hemisphere; only

you must not quite kill yourself in the doing of it.' The chair went to
H.H. Turner, chief assistant at Greenwich. Agnes Clerke commented:
'I hope the change will be fortunate for him. He will now at any rate get
the chance of showing what he has in him, which the Royal
Observatory had scarcely afforded him',[3] a remark that shows how
Agnes Clerke was influenced, perhaps unfairly, by Gill's critical
opinion of the Astronomer Royal.

Some time later (July 1894), in another catalogue of woes, Gill
revealed to Agnes Clerke, with full details (not recorded by his biographer) 'an intrigue against him worked through the Treasury'. It was, he
wrote, 'an item in the history of astronomy' and 'an example of what he
had to encounter in twelve years'.[4] Agnes Clerke sympathised:

> How paltry, how lamentable it all is! Must a man's life be spent in
> struggling for permission to do the work he is appointed to do? To me it
> seems one of the worst forms of injunction to cripple human faculty,
> and *such* faculty. As for Mr Christie (for I have no doubt that he is the
> one concerned) I can only hope that his motives are better than one
> would be tempted to suppose. One must try to believe and hope for the
> best. I am *very* grateful for your narrative of these unhappy
> complications. You will triumph over them with the patience and
> courage you are meeting them with . . . And in spite of these purblind
> departmental woes, you are preparing a grand record. I stand amazed at
> your list of voluminous and most important publications. You have
> indeed got ready a goodly supply of 'grist'.[5]

She gave him her practical support with an article on his heliome-
ter work in the next number of *Observatory*.[6] But even as her letter was
on the high seas, another arrived at the Cape with very cheerful news.

Frank McClean

Among the things that Gill had long campaigned for was a suitable
instrument for photography and spectroscopy. Upon the idea, first proposed in 1890, being 'coolly received' in London, Gill launched an
appeal for £5,000 at a public lecture in Capetown – but no benefactor
came forward.[7] Now, in 1894, Frank McClean, a wealthy and talented

engineer and amateur astronomer with a private observatory at Tunbridge Wells, wrote offering to donate a telescope (a 24-inch photographic refractor made by Grubb) to the Royal Observatory at the Cape.[8] Gill could not resist sharing the welcome (but not entirely unexpected) news with his friends.[9] He replied thanking his benefactor, and on the very same day wrote to his closest confidants – Agnes Clerke, Elkin and Kapteyn.[10] Agnes Clerke, 'in a tumult of joy', spread the news in a letter to the *Times* newspaper, published on October 1. The letter was 'very brief, but my sister strongly advised me to make it so, lest total exclusion should be the penalty of enthusiastic expansion'.[11] The Clerke sisters soon got to know the McClean family well. They stayed with them sometimes in their home in Tunbridge Wells where McClean still had his observatory, on one occasion being given a fine view of Saturn through his 10-inch telescope.[12]

McClean's telescope was to have spectroscopic facilities, and Agnes Clerke was not at a loss for ideas for its use.

> Three cheers for Mr McClean. He has supplied the greatest need of astrophysics at the present time. [She was referring to the opportunity of observing the spectra of southern stars.] My brain bubbles over with projects of what you will be able to do. Scores of questions present themselves to be answered by means of the 24-inch. It would be worth setting up for the sake of Gamma Argus alone! . . . From the time I first saw it I longed for a good photo of the ultraviolet spectrum, and now you are coming within reach of it, as well as so much besides. I shall be the torment of your life, 'wanting to know, you know' by every mail. It is a grand prospect.[13]

Agnes and her sister did not appreciate that telescopes are not made and erected overnight. Besides, there were delays and hitches concerning the design of the telescope and its dome. Gill's own quite definite ideas did not always coincide with the donor's. Gill was impatient to get matters moving; McClean took his time, while the maker, Howard Grubb of Dublin, tried to please them both. The story is recounted in detail by Ian Glass in his history of the Grubb family.[14] The McClean telescope, officially named the Victoria telescope in honour of Queen Victoria's Jubilee in 1897, was delivered in April 1898, though it was to be some years before it was in a satisfactory state for serious research.[15]

Agnes Clerke did indeed have 'scores of questions' that she wished to have answers for. Fascinated by the possibilities of spectroscopic radial velocity measurements, she tried doing her own calculations of masses and orbits of reputed double stars. Lady Huggins has recorded that in her later life Agnes Clerke took lessons in mathematics from her brother and derived much pleasure from them;[16] this was the occasion to which she referred. 'My mathematical efforts are moderately successful. I am creeping along, not flying, but with perseverance may learn enough to make me more competent as a critic. The work is very absorbing, but I do not allow it to become more than secondary', Agnes Clerke told Gill in 1894.[17]

Aware of many unanswered questions in stellar spectroscopy, she had begun a book 'experimentally'. Unlike her earlier ones in which she herself was the recorder of the work of others, this one sprang from her own ideas of what ought to be done and what might be done. 'It may not turn out well and in that case I will throw it to the winds.' In a fascinating description of a writer's inspiration she went on:

> But the idea has been so long haunting me that I feel driven to try if it could be realised. Have you ever read Dostoyeffski's [sic] tremendous 'Histoire d'un Crime'? He described how the mere thought of a murder got so solidified in a man's brain that at last he went and did it. Well, I often feel the same about the scheme of a book. It comes teasing and saying 'Write me'. And I reply 'the time has not yet come. I am busy about other things. Wait'. Then at last I find myself without further cause of postponement, and with a sigh and a plunge, I begin. And the beginning makes the end a deadly necessity.

The book in her mind was *Problems in Astrophysics*, not completed and published until almost a decade later.

An interesting sidelight to this story is that Agnes Clerke should have not only been familiar with the work of Fydor Dostoyevski but that she should have read Dostoyevski's famous novel in French as early as 1894. Dostoyevski was at that time practically unknown in England. The novel, *Crime and Punishment*, with its strange and violent character Raskolnikov, was published in 1866 in Russian and first revealed to western readers in a French translation. One wonders if she might have been introduced to Dostoyevski by Constance Garnett, Richard Garnett's daughter-in-law and famous translator of Russian literature including all of Dostoyevski's novels. Constance, then a young married woman, began translating from the Russian in 1893 (her Dostoyevski translations came later), and spent three months in Russia in 1894, the very year when Agnes Clerke found herself fascinated by Dostoyevski's creation.

Lick, fountain of research

Meanwhile, the topic of stellar spectroscopy dominated Agnes Clerke's correspondence with Holden at Lick, where William Wallace Campbell had begun his long series of superb photographic spectroscopic observations, including radial velocity measurements. Agnes Clerke quickly spotted his early paper on Nova Aurigae and commented on it to Holden.[18] She had not long to wait before Campbell himself sent her his further, unpublished, results, on the same subject (he published nine papers in *Astronomy and Astrophysics* between 1892 and 1894)[19] in which he reported changes in the Doppler velocity of the nova. 'I cannot thank you too heartily', wrote Agnes Clerke '. . . Congratulations on the success of your researches'.[20] Next time it was Holden who informed her of Campbell's work published in *Publications of the Astronomical Society of the Pacific*. The objects observed included Gamma Velorum, Agnes Clerke's favourite southern star.

> The achievement of observing a star at six degrees of altitude is of course unprecedented, and bears strong testimony to the extraordinary clearness of your climate, and to the skill with which so singularly disadvantageous [a situation] was neutralised. The results are astonishing. The simultaneous brilliancy of C [H alpha], and darkness of G, h and H [normal lines] seems altogether unaccountable.[21]

(The object is composed of two stars of types, in later terminology, WC7 and O7.) In 1895 Campbell communicated to Agnes Clerke his pre-publication results of the spectrum of the Trifid Nebula, another object which she had personally observed from the Cape, and his spectra of various other bright-line stars. In her letter of thanks she returned again to the subject of these latter stars 'which deserve the closest investigation; it is good news that you propose to follow this up'.[22] She gave publicity to the Lick work in a long report in *Observatory*.[23]

Campbell's colleague at Lick, John Schaeberle, also kept Agnes Clerke informed of his progress. There had earlier been a slight shadow between them on the matter of a reference in her *System of the Stars* which she ascribed to Holden, when in fact the work (a description of the Draco planetary nebula) had been jointly done by Holden and Schaeberle (a rare case of a mistake on Agnes Clerke's part). Schaeberle had protested, but was soothed by Agnes Clerke's apology.

> I regret extremely that anything in my book should have given you pain. Nothing was farther from my wishes or my thoughts than to show you the least discourtesy. But I unaccountably overlooked the fact that the paper quoted by me from the Monthly Notices was a joint production. Should the opportunity offer I shall be happy to rectify the mistake. My annoyance at its having occurred is all the greater that it concerns an astronomer from whom I entertain a sincere respect.[24]

True to her word, she amended the reference in the second edition of the book.[25] She thanked him for sending his papers and congratulated him on the success of his 1893 eclipse expedition.[26] In due course she received from Holden a glass copy of Schaeberle's eclipse photograph[27] and from Schaeberle himself a copy of his eclipse report.

Problems at Lick

Lick, however, had its problems during that period, where ever increasing tensions between Holden and his staff culminated in Holden's dismissal as Director in 1898. The long saga of Holden's troubles on many fronts and of how his authority was undermined by certain members of his staff has been narrated in detail by D.E. Osterbrock.[28]

There is no hint of these unhappy events in Agnes Clerke's correspondence with Holden; if she knew of them, she did not show it.
Holden was a cherished pen-friend of the Clerke family. In 1892, having
received and reviewed Ellen's little book, *Jupiter and his System*, he
took the trouble of sending her his detailed comments (which Agnes
'treasured up for her own guidance') as well as a booklet about Lick
Observatory with photographs of the staff.[29] 'We are all especially glad
of the likenesses in it', Ellen wrote, 'which enable us to realise you and
your staff as individuals. Of course your face was already, to us, like that
of a friend, both from the many published portraits and photographs, as
well as from that in my sister's possession.' When the poet and writer
Sir Edwin Arnold published an account of a reading tour of America
that included a visit to Mount Hamilton and a meeting with Holden
and his son, Agnes declared that the son 'will be very perverse if he do
not take up astronomy and help to found a dynasty of Directors at Lick,
as the Struves have done at Pulkovo'.[30] This, alas, was far from being a
possibility. Though Agnes Clerke cannot have known it, Holden was
separated from his wife, and his son Edward was with his father only
during summer holidays.[31]

Holden did little astronomical research in his last few years but
spent his spare time writing on topics of special interest to him, not all
of them scientific. He was a man of wide cultural tastes, which he could
share with the like-minded Agnes Clerke. Among his gifts to her over
the years were an article of his on Balzac (1888) ('read with great interest
and surprise to find what a profound and persistent study you had made
of a subject so remote from the more absorbing pursuits of your life'), his
'admirable translation' of an essay by Sully Prudhomme (1891)[32] and an
article on West Point Academy. When preparing a book on *The Mogul
Emperors of Hindustan* (1895) he appealed to her for help in identifying
certain oriental personages from pictures. Agnes Clerke and her sister
went to a great deal of trouble in an effort to solve the problem, consulting with authorities in the British Museum and with the Professor of
Persian at the Collège de France.[33] Agnes Clerke received a copy of the
beautiful book when it appeared. 'The very looking over it gives pleasure and instruction',[34] she wrote.

Among scientific publications which Holden sent to Agnes

Clerke during his last months in office were his photographic *Lunar Atlas*, his *History of American Astronomy*, and his splendid essay on Mountain Observatories, for all of which she expressed her admiration and appreciation.[35] It is a tribute to Holden's stoicism that he could trouble to send these gifts while under the shadow of his departure, not just from the observatory but from astronomy itself: his decision to give up had been taken in May, and he left Mount Hamilton forever on 18 September 1897.

Agnes Clerke's last letter to Holden – a postcard dated 12 July 1897 – ends the correspondence with him preserved at Lick. It reads: 'Many thanks for your valuable article on the history of American Astronomy which I have read with admiration of the lucidity and completeness of treatment given to that important subject.'[36] There is no hint that she knew of his impending resignation.

It is impossible to imagine that she, who owed so much to Holden in the early days of her career, would not have kept in touch with him afterwards. Holden eventually made another life for himself as librarian at West Point Academy; however, the archives there[37] contain only his official papers. He died in 1914, aged 67.

Harvard

Agnes Clerke's contacts with Harvard, that other mighty centre of astrophysics in the United States, was somewhat formal by comparison with her highly personal relations with Lick and the Cape. In 1897, however, she made bold to approach E.C. Pickering at Harvard directly with a specific request. Being particularly fascinated by variable stars, she felt that more could be done about testing them for duplicity. She had just published a paper entitled 'A new class of variable stars' in the *Observatory*,[38] which dealt with stars having short-period symmetrical fluctuation in light, in which the intervals from minimum to maximum and from maximum to minimum were almost exactly equal. She predicted that spectroscopic observations would show these to be binary stars in contact with each other. She sent a copy of her paper to Pickering politely venturing to call his attention to her suggestion.

If it be correct, the variables U Pegasi and S Antliae possess double
spectra, the lines in which are relatively displaced to a very large extent
at the maxima of the stars. With your powerful apparatus, the
phenomenon might be discoverable. Perhaps you will think it
worthwhile to have some test photographs taken?[39]

The stars in question, Agnes Clerke believed, were similar to
Beta Lyrae, a star long-known for its periodically varying brightness. In
1895 Antonia Maury at Harvard had observed that the spectrum, too, of
this star varied periodically. Agnes Clerke believed that the same observation would yield a similar result in the other cases that she mentioned. These stars could be interpreted as eclipsing pairs but with the
component stars very close together.

The letter was acknowledged by Pickering[40] who had a week or
two earlier dispatched to her some photographs of the Argo Nebula (Eta
Carinae) taken with the Harvard Observatory Bruce telescope in
Arequipa. Agnes Clerke was most appreciative of these ('Permit me to
express my admiration for the splendid photographs'.)[41] Pickering did
not, however, to her great disappointment, make the observations
which she had asked for. Her request reveals how far removed Agnes
Clerke, sitting in her basement study in Redcliffe Square, was from the
reality of astronomical life. The output of the Harvard astronomers was
nothing short of astounding; yet Agnes Clerke seemed to imagine that a
set of light and radial velocity curves could be ordered like items on a
grocer's list.

Some months after this fruitless request to Pickering, we find
Agnes Clerke writing to him to introduce a friend who was visiting the
United States. She was Lady Margaret Domvile, sister of the Earl of
Howth, a prominent Irish nobleman.

> She is a person of distinguished ability, and takes a deep interest in
> everything that concerns intellectual progress. Naturally, then, she
> would be loth to quit Boston without seeing something of Harvard
> College and its world-famous Observatory, and I confidently apply to
> your kindness to procure for her the requisite facilities. She would be
> particularly pleased, I feel sure, to meet your staff of ladies, and hear
> from Mrs Fleming some particulars as to the nature of their work.[42]

Mrs Williamina Fleming, leader of the team of women employed on the Draper memorial Catalogue of stellar spectra at Harvard College Observatory, had been assigned the task of developing an empirical classification system for the more than 10,000 stars which made up that great catalogue published in 1890. In the course of this work she had discovered large numbers of objects with unusual spectra, including novae and variable stars. At the time when Agnes Clerke sought the introduction for her friend, Mrs Fleming, already famous, was very much in the news on account of her recent discoveries in the Magellanic Clouds of hundreds of similarly interesting stars, among them stars with emission-line spectra.

Lady Domvile was duly received at the Observatory at Harvard,[43] and presumably met Mrs Fleming and her team of ladies.

A new era at Lick

When Holden left, John Schaeberle was placed temporarily in charge of the Observatory. Agnes Clerke was quick to consolidate her Lick contacts by writing to Schaeberle complimenting the observatory on the magnificent set of lunar photographs, and congratulating him 'on the distinction of being selected to direct *ad interim* the establishment on Mount Hamilton which may be taken as an augury of future definitive promotion'.[44] As regarded this prospect, however, Schaeberle was disappointed, since the post of director went to James Keeler the following year (April 1898).

Keeler, whose brilliant but short career has been chronicled by his twentieth-century successor at Lick, Donald E. Osterbrock,[45] was recognised by his contemporaries at the end of the nineteenth century as America's leading astronomical spectroscopist. He was one of the earliest students at Johns Hopkins University in Baltimore, the country's first research university, and had spent a year in Germany studying at the feet of that country's illustrious physicists, von Helmholtz, Bunsen and Kayser. In 1886 at the age 29 he joined the staff of Lick Observatory under Holden, who was a warm supporter, and in the following years established his reputation with his observations of emission-line stars. This work was included by Agnes Clerke in the chapter on such objects in *The System of the Stars*. From Lick he was appointed Director of Allegheny Observatory in 1891 where in 1895 he made the observation for which he is popularly famous, the photographic spectrum of Saturn which beautifully revealed orbital motions in the rings. Now in 1898 he was back in Lick as its second Director.

Agnes Clerke already knew Keeler, whom she had met in London two years previously. Accompanied by his wife, Keeler had come to Britain to attend the Liverpool meeting of the British Association for the Advancement of Science in August 1896, where he gave an invited lecture on the spectroscopy of planets.[46] After the meeting he visited London and was entertained at their home by the Clerkes.[47] He also on that occasion visited the Hugginses at Tulse Hill and met Gill, who happened to be in London at the time.

Before meeting him in person, however, Agnes Clerke had long been in correspondence with Keeler. When Lockyer's meteoritic hypothesis was under discussion, just after the Hugginses had published their wavelengths of the emission lines in the spectrum of the Orion Nebula, Agnes Clerke wrote to Holden with a suggestion.

> Since you are making spectroscopic observations, I cannot resist mentioning a point which I have for some time wished to see settled, and which can no-where be settled if not at Lick. It is, whether or not there are bright lines to be seen in the spectra of variables of Secchi's fourth type. The interest of the question lies in its bearing on Mr Lockyer's classification of the 'celestial species'.[48]

The type mentioned [later identified as carbon stars] were, in Lockyer's scheme, at the end of their life-cycle and (in Agnes Clerke's words) 'semi-extinct'. Bright lines in such spectra would disprove Lockyer's hypothesis, she claimed. She enclosed a list of candidates which Holden passed on to Keeler with every intention of trying.[49] 'I feel it a singular privilege to be allowed to put questions to your great telescope',[50] she wrote.

Nothing more is mentioned in correspondence about this programme, but Keeler was in fact obtaining other spectra which were highly relevant to the Lockyer debate. These were his spectra of the

Orion Nebula, already mentioned, which confirmed – indeed improved on – those of the Hugginses and clinched the argument against Lockyer's interpretation of the origin of the nebular lines. Aware of the value of publicity in the academic world, Keeler announced his important results in advance of publication to several leading astronomers – and also to Agnes Clerke,[51] who was now 'the chief astronomical writer in the English-speaking world' and 'an important opinion moulder'.[52] Her reports in Britain on American work were often reprinted in America. She had recorded and backed the Hugginses' Orion finding in *The System of the Stars*, but Keeler's result had come too late to be included in the book.

Keeler kept her informed of his progress on the spectroscopy of the Orion and other nebulae, and of Beta Aurigae, an object in which Agnes Clerke had a particular interest.[53] His fundamental paper on nebular spectra, written and published while he was at Allegheny, drew congratulations from Agnes Clerke, to whom he had sent a copy.[54] On his appointment to the Directorship of Lick, Agnes Clerke sent congratulations from her sister and herself 'on the brilliant promise of a fresh career at Lick'. He had resumed his spectroscopy of the Orion Nebula, and Agnes Clerke made some interesting comments on the fact that the image of the nebula was more extensive in hydrogen light than in the light of the mystery element 'nebulium', 'suggesting the inference that nebulium is a heavier stuff than hydrogen', but that 'until nebulium is captured we shall not know the truth'.[55]

When in 1899 Keeler made his important discovery of the spiral shape of innumerable small nebulae on long-exposure photographs taken with the Crossley reflector, he hastened to inform Agnes Clerke and sent her slides of several of his spirals as well as some large transparencies which he asked her to have a look at.[56] Some were of well-known spirals already photographed by Roberts; others were newly recognised. 'My work with the Crossley reflector has led me to the belief that practically all small isolated nebulae are spirals though with exceptions (the Lyra nebula and the Dumbbell nebula, for instance)', he wrote. He found that all the 'pretty bright, round, brighter in the middle' nebulae, turned out to be spirals like M101, and the 'much elongated' ones were spirals seen edgeways. That the great majority of nebulae were spirals seemed to him very interesting and important, but, with his customary caution, he intended to wait for more data before publishing anything. He continued: 'Your paragraph on p. 265 of *The System of the Stars* contains a prophecy which seems in a fair way of being verified.' Keeler was referring to Agnes Clerke's comment that 'curved furrows of light such as is agreed to call 'spiral' have been traced in many planetary and annular nebulae; with still greater optical power than is now at the disposal of astronomers, they might possibly be brought into view in all'.[57]

Agnes Clerke was overwhelmed by Keeler's photographs.

> No words can be too strong to express my admiration of these wonderful pictures. The definiteness with which they bring out what may be called the 'law of spirality', as well as some of its conditions, is amazing. So far as I can perceive, there is always a double origin to the spirals, one set issues from one extremity of a diameter of the nucleus, one from the opposite. . . . Another highly significant feature is that the nebulous streamers are deflected by the stars they approach. This of course shows that the stars, far from having condensed out of the material of the streamers, existed antecedently to their formation. Your generalisation regarding the spiral structure of most nebulae is of profound importance, and I am greatly pleased and flattered that you should remember my imperfect forecast of it.[58]

This exchange of letters shows Agnes Clerke as not just a publiciser of others' work but an active participant in discussions about its interpretation. She was right about the typical shape of spiral galaxies, but quite wrong in her picture of gas streaming away from stars.

Agnes Clerke's last letter to Keeler was dated 14 August 1900, two days after his unexpected death. His colleague W.W. Campbell, replying on behalf of Mrs Keeler, described Keeler's final illness and gave a positive answer to Agnes Clerke's request for Lick material to illustrate a future book.[59] Agnes Clerke responded with a moving letter of condolence:

> Our remembrance of him was of a man, not only young, but likely to remain young almost indefinitely. He seemed so alert, so well-balanced, so eminently sound in mind and body. He has left a fine

> record of work done, but one hoped that it would have shrunk into
> insignificance by comparison with all that he *would* accomplish. And
> now the column is truncated! We do not see the reason for it now, but
> we shall some day. After all, the very longest life is only a beginning;
> the ravelled threads will be knit together beyond the grave.[60]

She then expressed the wish that Campbell, who was temporarily in charge, would succeed as director, as 'any other choice would
seem unfortunate'.

Campbell, supported by all the leading American astronomers,
was indeed appointed director in December 1900, and continued to be
an obliging friend to Agnes Clerke, whom he knew since his visit to
London two years earlier on his way back from the Indian eclipse. On
that occasion Agnes Clerke had told Gill how in Campbell's company
'time fell short sooner than materials for conversation'.[61] On his being
formally appointed she wrote: 'I rejoice to see that my wishes have been
realised by your appointment as Director of Lick Observatory – the most
enviable post in the world for an astronomer like yourself endowed with
the ability and determination to enlarge to the uttermost the bounds of
science',[62] ending her letter with love from her sister.

Mira Ceti

In February 1899, Agnes Clerke wrote to Campbell, who with the
Mills spectrograph acquired in 1896 was a world leader in the field of
radial velocity measurements, congratulating him on his spectroscopic observations of the variable star Mira Ceti. She had reported on
his work on four successive occasions in the *Observatory* during that
year and continued to report on it regularly thereafter. The star showed
emission lines apparently split into three components. She imagined
this to be the Zeeman effect in action (the effect, discovered by Pieter
Zeeman in 1897, showed itself in the splitting of spectral lines under
the influence of strong magnetic fields). She felt sure (but without
giving reasons) that here lay 'the key to the enigma of variability'.

> I am most anxious that you should try a simple experiment on the
> subject, the success of which would go near to placing that key in your

hands. The experiment consists in applying polariscopic tests to the tripled rays. If they are of magnetic production the lateral lines should be oppositely polarised from the middle line. Consequently on the interposition of a Nicol they ought to disappear and reappear in alternate positions from the leader. But if they are circularly polarised, I need scarcely say that a plate of mica or some equivalent should be added to the apparatus.

She had, she informed him, earlier published this suggestion in the *Observatory* as a test for the spectrum of Beta Lyrae but it had not been acted on by anyone. 'The next ensuing maximum of Mira will then perhaps be celebrated by the discovery – and you are evidently meant to be the discoverer – that its light changes are of magnetic origin.' Campbell did in fact try the experiment – which cannot have been an easy one – but without a significant result. 'I was not unprepared', she wrote, 'for the perverse behaviour of Mira Ceti. It seemed as if the star out of "pure cussedness" determined to baffle your efforts to put it to a severer test than had ever been applied before'.[63] (This may well have been the first ever attempt to observe the Zeeman effect in astronomy; the first recorded instance is Hale's observations of the effect in sunspots, made in 1908.) Campbell did not publish his ambiguous result on Mira, but Agnes Clerke reported it in her *Problems in Astrophysics*, then in preparation, saying that at its maximum brightness in 1899 'the star unfortunately failed to replenish its due measure of light, and gave an imperfectly legible spectrum'.[64] The change in the spectrum of Mira was eventually found to be due to pulsation.

Nova Persei

On 22 February 1901, the discovery of a naked-eye nova, Nova Persei, was announced. Agnes Clerke lost no time in contacting Campbell at Lick, spurring him to action. 'Doubtless you have already made important observations of Nova Persei. Have you thought of testing it for polarisation effect? The opportunity seems a magnificent one for getting a definitive answer to the question, vital to the progress of astrophysics, whether they occur in such objects', she wrote on 28

February.[65] Clearly, she was still thinking of the Zeeman effect. Campbell obtained successful spectra of the nova, which Agnes Clerke immediately reported in the *Observatory*,[66] but there is no indication that he tried out the polarisation experiment. A week later she wrote to Pickering – whose award of the gold medal of the Royal Astronomical Society had been announced – congratulating him on the Harvard observations showing changes in brightness and spectrum of the nova, but (being apparently less at ease with the Harvard astronomer than with her Lick friends) made no request regarding polarisation.[67] She was disappointed that no-one was interested in her idea. 'I wish a patient investigator would take up the question of the spectra of new stars', she wrote to Gill.[68]

Research ambitions

By degrees, Agnes Clerke had become not just a recorder of astronomical progress but had ambitions to be a player in the research itself. Learning from McClean that his new telescope was almost ready for use, Agnes wrote to Gill:

> Soon I shall be looking for results and answers to my curious and impertinent questions about spectra of nebulae etc. I think, by the way, that you might get an important result and *wholly novel* by taking a spectral photo of Omega Centauri or 47 Toucanae, I mean by turning a condensed image of the cluster on the slit so as to get the integrated light of the crowded stars. I want you to announce a 'discovery' straight off, so here is your chance![69]

It was a good idea – a way of discovering the overall spectral type of stars in unresolved globular clusters. There is no record that her wish was carried out, though Gill obtained a beautiful direct photograph of Omega Centauri. When Gill told her he was planning to observe some of her favourite stars (the objects she had observed at the Cape many years before) she replied 'It is very good news that Eta and Gamma Argus are not to escape your potent attacks, and I feel confident the results will not disappoint my ardent anticipations'.[70] But she wanted more: 'I do not know whether you care to pick out spectroscopic binaries, but I venture to suggest that Eta Argus [i.e. Eta Carinae] should be placed on your list at the first opportunity.' Gill published observations of the spectrum of this object in 1901.

She continued to exhort Gill, naming stars whose spectra she wished to see photographed, and impatient at the slow progress of observational astronomy while the solutions to so many problems were, she thought, within reach. On one occasion she seemed to reprove her friend for his involvement in other programmes. This somewhat offended him, but she explained herself in her usual tactful tone:

> You must not be surprised at different people appreciating only a part of your work. It only shows how wide its scope is. Nor am I going to take lying down that you imply a want of sympathy on my part for the Durchmusterung . . . I protest emphatically that I welcomed the parallax performances quite as warmly. Indeed only want of time restrained my pen from discussing them.[71]

Agnes Clerke was alluding here to the monumental account of the Cape work published in the *Annals of the Cape Observatory*, which she had looked forward to using for the revised *System of the Stars* then in preparation. It included the results of the parallaxes of minor planets, the Durchmusterung catalogue, and Kapteyn's analysis of the Durchmusterung material with his conclusions as to the distribution of stars of various types in the sky. Kapteyn did not have long to wait before receiving the stylish Clerke treatment in an article on the sun's motion which also embraced the recent work of Newcomb and Campbell. It ended with speculation on our future journey through space.

> Each year, accordingly, we explore a belt of space nearly 400 millions of miles in width, and our travelling, like that of the clouds, is 'irrevocable'. Shall we find ourselves in an 'ampler ether' as we proceed? Or will the wreckage of our little planet help stock the void with meteorites?
>
> *It may be that the gulfs will wash us down:*
> *It may be we shall touch the Happy Isles.*[72]
>
> Even the poets scarcely knew for certain which fate overtook Ulysses when he 'sailed beyond the sunset' into a newer world.

11 Popularisation, cryogenics and evolution

Astronomy 1898

Among Agnes Clerke's activities during the last years of the century was the part authorship of a book entitled simply *Astronomy*, one of the Concise Knowledge Library series.[1] Her co-authors were the astronomical spectroscopist Alfred Fowler, Director of the Astrophysical Laboratory at the Royal College of Science in London, whom she knew of old through Lockyer, and John Ellard Gore, a well-known Irish amateur astronomer. This book was Agnes Clerke's single excursion into 'ordinary' popular writing. Her contribution comprised the sections on the history of astronomy and the solar system. The reviewer in *Nature* wondered why 'such a formidable array of authors' was needed when any one of the three could have written it alone.[2] There was needless repetition and nothing for 'the serious student' that was not already available in Ball's popular books. One assumes that the publisher wanted top names and that the authors dashed off their contributions without much consultation.

Robert Stawell Ball and Gore were the leading popularisers of astronomy in the English language at that period, the successors of Richard Proctor, who died in 1888. Writing and lecturing had been Proctor's only source of income. Ball's circumstances were entirely different. He held chairs of Astronomy first at Trinity College Dublin and then at Cambridge. He was a classical astronomer and a first-rate mathematician whose contributions to science were his books on mechanics and his university textbook on spherical astronomy. Yet public lecturing and writing occupied a very large part of his life and of his energies. Having discovered his talent for public speaking he organised his diary around lecture engagements throughout Britain, Ireland, or wherever he found himself on business, including Canada and the United States. To diverse audiences he gave Royal Institution discourses, the Gilchrist Trust lectures (an adult education fund), addressed official town hall functions, workers' clubs, even prison inmates. He never lectured without a fee and thus earned a substantial extra income, boosted still further by his enjoyable popular books which were often bi-products of his lectures. He is reputed to have been heard by a million people in the course of his lifetime. In his defence it has been said that in this way 'he was able to produce material that assisted very considerably towards the scientific education of the population of the British Isles over five decades from 1870 into the 1920s'.[3] Of the 22 books he published, 13 were purely popular. Ball took little interest in astrophysics: his compendium of short biographies of famous astronomers, *Great Astronomers*, ends with John Couch Adams. The titles of his best-loved books – *In Starry Realms, Story of the Heavens* – give a clue to their style. A young astronomical journalist, Hector MacPherson, described him as 'one of our highest authorities in speculative astronomy', this being intended as a compliment.[4]

Gore was an equally productive but less flamboyant populariser. An engineer by profession, he had developed an intense interest in astronomy while working in India and had retired early on a pension in order to devote himself to it full-time. He was a tireless and talented observer who, with no more than a 3-inch refractor in the unfavourable climate of Dublin, discovered several variable stars. He published catalogues of variables and of binary stars and acquired a reputation as a writer of popular books and articles. He was active in amateur astronomical affairs, first in the Liverpool Astronomical Society and later in the British Astronomical Association.

Gore contributed many simple and informative articles to the magazine *Knowledge*. His favourite topic was the stellar system, about which he did some research from published statistics. He wrote five or six popular books aimed at the non-scientific reader, with titles such as *Scenery of the Heavens, The Worlds of Space* and *The Visible Universe*. He favoured a finite universe, as Agnes Clerke did, but speculated on other universes beyond our observational reach, and on the possibility of life on other planets. He had, so MacPherson recorded, 'a noble appreciation of the vastness of the universe, and an earnest desire to reach the truth'.[5] *Nature*, however, in a harsh mood, described a book of Gore's as 'disconnected essays' and found Ball's work 'scrappy'[6] – not surprising considering the speed at which they were produced.

Another favourite author of the speculative variety was the French astronomer and balloonist Camille Flammarion. Flammarion, born in 1842 (thus the same age as Agnes Clerke) began his career as a young assistant under the tyrannical Leverrier at the Paris Observatory but threw off the shackles (as he saw them) of classical astronomy and turned to lecturing and writing on exciting descriptive subjects.[7] He set up his own private observatory at Juvissy near Paris specifically for observing the planets and for popular teaching. The French Astronomical Society, which he founded in 1877 and which still flourishes, was a model for other similar groups and inspired many of the outstanding French astronomers of the twentieth century.[8] Flammarion was the most successful populariser of astronomy in his lifetime. His books *Mars* and *Other Worlds*, and his famous *Popular Astronomy* (first published in 1880) were translated into English, the last by Gore in 1897.[9] In the preface to this translation, done 'with the author's sanction', Gore records that 'no fewer than 100,000 copies of this work [were] sold in a few years – a sale probably unequalled among scientific books'; it received a prize from the French Academy and was selected by the Ministry of Education for public libraries. Gore updated the English version with extra illustrations and with interpolations of his own in brackets.

The names of these three writers – Ball, Gore and Flammarion – appear briefly in Agnes Clerke's books. Their own books, however, are purely popular, in no way comparable to hers. Her purpose was entirely different and she had no rival.

The Huggins *Atlas* 1899

Sir William and Lady Huggins' beautiful *Atlas of Representative Stellar Spectra*,[10] the result of a decade of labour, gave Agnes Clerke an opportunity to praise her friends' achievements in 'an observatory half-shadowed by London smoke fogs'.[11] It comprised a set of fine reproductions of spectrograms of bright stars, and a text giving the history of their spectroscopic programme illustrated by Margaret's artistic woodcuts. The *Atlas* was clearly intended (by Margaret?) to put the Huggins' work on record for posterity. Though Agnes Clerke would not dream of saying so, the fewness of the spectra underlined the inefficiency of a city site in the new era of mountaintop astronomy. Agnes Clerke's review concentrated on the question of the evolution of stars without entering into controversy or mentioning names, and is interesting for her appreciation that surface gravity, as well as temperature and pressure, is likely to be a factor. The Huggins *Atlas* earned for the authors – perhaps more for its artistic presentation than for its content – the Actonian Prize of the Royal Institution, the same prize that had been awarded to Agnes Clerke three years previously.

Agnes Clerke was clearly personally very attached to the Hugginses, and never lost an opportunity for singing their praises. An example is her effort to have William Huggins' name immortalised in spectroscopy by having the hydrogen Balmer lines named after him. In 1879, Huggins photographed the ultraviolet spectrum of Vega, which showed what was clearly a continuation of the series of five hydrogen lines in the visible part of the spectrum, beginning with the red line (H alpha) then known as C. Balmer produced the series in the laboratory in 1885 and also worked out the formula describing the pattern of the lines. It was thus given the name 'Balmer series'. Agnes Clerke attempted to have Huggins' name associated with the series (one wonders how far she was prompted by Margaret), and in *The System of the Stars* (1893) she labels the ultraviolet lines 'Huggins alpha, beta, gamma etc'. In the second edition of 1905 she designates the series the 'Huggins series'. The description, however, did not catch on: the series remained Balmer's.

A pet ambition of Huggins, which occupied him for years, was to photograph the solar corona without a total eclipse. He made several attempts, beginning in 1882, at the Tulse Hill Observatory, and later recruited others to try on his behalf and under his strict instructions at better astronomical sites overseas. Agnes Clerke gave extensive coverage in her *History* to Huggins' persistent efforts, which continued for a decade or more;[12] but she had to admit in 1905 that the prospect of succeeding was dim. It remained, she added rightly, 'a prime desideratum in solar physics'.[13]

The Hodgkins Trust Essay, 1901

Agnes Clerke's award of the Actonian Prize in 1893, which made her a recognised figure at the Royal Institution, coincided with the start of researches into the liquefaction of gases at the Royal Institution, for which Sir James Dewar became widely known. Dewar was the inventor of the vacuum flask named after him with which, in 1894, he succeeded in freezing air. Agnes Clerke took a keen interest in Dewar's work and attended all his lectures and demonstrations. She recorded excitedly to Gill in May 1898 that 'after five years of trial and failure Professor Dewar has produced palpable undeniable liquid hydrogen'[14] – for the first time. The following year, as part of the centenary celebrations of the Royal Institution, Dewar planned a public demonstration of the liquefaction of hydrogen. 'Professor Dewar is in despair, he says, of being able to produce liquid hydrogen before his audience on the 7th [June 1899], but his "despair" is perhaps only an augury of success',[15] wrote Agnes Clerke. She was right: the experiment worked and the audience in the famous lecture theatre of the Royal Institution witnessed a thrilling display of this 'preternatural substance', which shortly afterwards he also solidified. Dewar's next public triumph was the solidification of oxygen in front of a crowd audience in the same lecture theatre in April 1900.

Dewar's experiments required massive and enormously costly refrigeration machinery, much of it provided out of a large sum of money donated to the Royal Institution in 1895 by an American benefactor, Thomas G. Hodgkins, for the support of 'investigation of the relations and co-relations existing between man and his Creator' – a gift very much in the spirit of the Actonian benefaction. In addition, in recognition of his successes, Dewar became in 1899 the first recipient of the Hodgkins gold medal of the Smithsonian Institution of Washington. It was decided that the cryogenic researches at the Royal

Institution which flowed from this benefaction should be placed on record in the form of a Hodgkins Trust Essay. Agnes Clerke was invited to do this. Her eighteen page essay, entitled 'Low Temperature Research at the Royal Institution 1893–1900',[16] and illustrated with photographs of Dewar's historic apparatus, reviewed the history of the subject with particular reference to Dewar's contributions (in about 70 publications) and discussed the chemistry and physics of matter at low temperatures. It also included an account of the recent discoveries of the inert gases. The essay, in a field quite outside that with which she is normally associated, is an impressive example of Agnes Clerke's versatility and application. Agnes Clerke was elected a Member of the Royal Institution on 3 March 1902, her proposer being James Dewar.

The University of Glasgow celebrated its 400th anniversary Jubilee in June 1901. Agnes Clerke received a formal invitation (printed in Latin, with the pronouns given a feminine gender by hand). It was an indication of her increased reputation, though, true to her bashful nature, she does not appear to have attended.

The on-going *Dictionary of National Biography*

An enduring legacy of Agnes Clerke is her contribution to the *Dictionary of National Biography*.[17] The Dictionary, begun in 1882, was a compilation of short biographies intended to include every person of interest born in Britain and Ireland in recorded history (the living excluded). The original dictionary with over 29,000 entries appeared volume by volume in alphabetical order at the rate of one per quarter from 1885 to 1900 with a supplement to 1901. The bulk of the work was done by a core of 100 regular contributors, of whom Agnes Clerke was one (and the only woman), with responsibility for the biographies of astronomers and allied scientists.

The 150 subjects of Agnes Clerke's thoroughly researched biographies include the Astronomers Royal, the three Herschels and almost every other astronomer of note as well as many of lesser fame. Mathematicians such as Charles Babbage, chemists such as John Dalton and Edward Boyle, opticians and instrument-makers such as the Dollonds and T. Grubb, are to be found among her entries.[18] Ellen Clerke also contributed some biographies, three of them of scientific interest – Richard Proctor, Mary Somerville and Mary's husband William Somerville. It would appear that in the early stages, writing for the *DNB* was to be a family activity in the Clerke household, as there is one entry (of George Barrett, an actuary) signed JWC (their father John William Clerke). It is rather pathetic to note that this single entry in volume I of the *Dictionary* represents his whole effort (he died in 1890). The identity of the writer was a puzzle to later editors whose records showed that the entry was sent in by Miss E.M. Clerke.[19]

History, fourth and last edition, 1902

A new edition of the *History of Astronomy during the Nineteenth Century* was published in 1902. This was the fourth and last edition, which was reprinted in 1908 after Agnes Clerke's death. This edition, revised and corrected, has the same chapter contents and headings as the third edition of 1893, and the same illustrations, with the sole addition of a photograph by Gill of the comet of 1901. The chronological table is updated to 1901, the last recorded event being the unveiling of the McClean 'Victoria' telescope at the Cape. The list of the world's large telescopes includes the 60-inch at Yerkes Observatory installed in 1902.

Man in the Universe: Alfred Russel Wallace, 1903

Though not discussed in her major books, the weighty subject of biological evolution attracted Agnes Clerke's interest in the early years of the twentieth century. Her particular involvement was in the cosmological aspect of evolution as propounded by Alfred Russel Wallace, cofounder with Charles Darwin of the theory of natural selection. A book by Wallace, *Man's Place in the Universe*,[20] published in 1903, argued on physical as well as biological grounds that human evolution was unique.

In the debate on the 'plurality of worlds', the majority accepted view, throughout most of the nineteenth century, was in favour of many such other abodes of life. Wallace decided to examine the scientific grounds for this view. His arguments were recalled by the cosmologist Frank J. Tipler in 1981, who recognised them as 'exactly the same as those given by modern evolutionists. Thus the biological arguments against the evolution of intelligence [elsewhere in the universe] have not changed in 75 years. The great evolutionists have always been united against ETI (extra terrestrial intelligence).'[21]

Wallace's highly readable book set forth all the evidence available. It began with the history of astronomical discovery up to the most recent observations about the nature of stars and their distribution in space. This entailed establishing, as far as it was possible, the position of the sun among the multitude of stars and nebulae, and the sun's movement in space. Then followed the question of the likely existence of other planetary systems (based on the observed proliferation of double stars), the chances of suitable physical conditions on other planets, and, finally, the possibility of the evolution of human life on these worlds. Agnes Clerke's *History of Astronomy during the Nineteenth Century* and *System of the Stars* were among Wallace's astronomical authorities, but he also consulted her directly, as emerges from her letters to him, written between 1901 and the time of publication of the book. 'It is most gratifying to me to hear that you have read my books with approval. Words of recommendation from such a source are not easily forgotten', Agnes Clerke wrote when she replied to his first queries, about how Copernicus and Kepler estimated the distances of the planets, and about the meaning of 'annual parallax'.[22] She later sent him, 'on the chance of its being useful to you', a paper by Newcomb with his best datum on estimating stellar distances statistically from proper motion ('secular parallax').[23] Another question put to Agnes Clerke was whether the sun would proceed indefinitely in its present direction in space (something which would affect the sun's apparent position near the centre of the stellar system), to which she replied that 'there is not the slightest warrant' for it. She was also asked about the 'solar cluster', a group of nearby stars, postulated by the South American-based astronomer Benjamin Gould, to which the sun itself belongs. She had subscribed to the reality of the solar cluster in the *System of the Stars* (first edition), but was now quite cautious about it, saying that the work of Kapteyn, 'the foremost investigator of sidereal construction', had not confirmed its existence.[24] (The second edition of *System of the Stars*, two years later, was to voice that caution: 'The grounds upon which it was based have been nevertheless gradually undermined by close and varied research.')[25] This was a disappointment for Wallace; the position of the sun within a cluster, itself in the centre of the stellar system, was one of the supports of his theory of the uniqueness of the earth's position in the universe. He sent Agnes Clerke the relevant chapter of his typescript for her comments. Her only criticism related to the solar cluster which, she told him, 'has dropped out of Professor Kapteyn's scheme of the Universe'.

> One must not be surprised where such enormously complex masses of data have to be treated, to find that conclusions regarding them have to be revised. Accept my apologies on behalf of astronomical science which has no more laborious and candid notary than Kapteyn of Groningen.[26]

Wallace, on balance, considered that the solar cluster was real, but recorded Agnes Clerke's reservations in his book. His principal contention, based on biological–evolutionary considerations, did not depend, of course, on the existence of the solar cluster. In fact, the idea of a solar cluster survived as 'Gould's Belt', but not as the tight group of stars envisaged by Wallace as occupying the centre of the Milky Way System. Wallace also sent Agnes Clerke, for her comments, the correspondence he had with George Darwin, the Cambridge mathematician famous for his theory of the tides, on the question of the stability of the Milky Way. Darwin had passed Wallace's queries on to the brilliant young mathematician Edmund Whittaker, then secretary of the Royal Astronomical Society. Whittaker was working on the idea that the principal forces involved might be electrodynamic rather than gravitational. Agnes Clerke thanked Wallace for the privilege of perusing the letters.

> Mr Whittaker is a mathematician of the very highest order. Perhaps not more than a half a dozen men have lived in this century back who have

rivalled his amazing intuitive power. So that one listens almost with bated breath to the subversive views he expresses in his letter to Professor Darwin. As to the spiral nebulae, I have long felt sure that they were dominated by forces other than gravitational though one cannot doubt that gravity exercises its power besides.[27]

The discussion made up the last chapter of Wallace's book, in which he left the matter open.

Wallace's ideas, when originally published in an article in a periodical, were immediately attacked by astronomers, notably in a review by the opinionated Maunder in *Knowledge* who spoke of 'myth' and 'speculation'.[28] The magazine's editor thought the controversy might be prolonged to its advantage and asked Agnes Clerke 'to keep the ball flying'.[29] She sportingly obliged, with a letter published in the next number of *Knowledge* dealing with the principal astronomical area of contention, namely the solar cluster with the sun at its centre. She pointed out that whether or not the sun was at the centre made no difference to the main issue, and repeated her usual arguments for a limited universe.[30] The controversy continued. The French populariser Camille Flammarion put the atheistic view – that 'theologians refuse to admit the infinity of space so as not to give it an attribute of God'.[31]

Agnes Clerke received a copy of *Man's Place in the Universe* as soon as it came out.

It is very kind of you to have your remarkable book sent to me. I value it greatly and hope to study it attentively. It is unique from the breadth of knowledge brought to bear upon a subject of peculiar interest. No great biologist has ever before seriously considered the possibilities of cosmic life; and they can only be fitly discussed in the light of expert biological science.[32]

Agnes Clerke made Wallace's book and its implications the subject of an article in the *Edinburgh Review* in 1904.[33] In that unsigned article she reveals herself as very impressed by Wallace's arguments. His picture of the universe coincided with hers, one whose entire contents, of stars, nebulae, and everything else, form one single system. 'We cannot but agree with Dr Wallace that the outcome of recent investigations . . . is entirely opposed to the old idea that countless myriads of stars *all* had planets.' She did not, however, rule out the possibility of life elsewhere in the universe. 'Our thoughts are, nevertheless, to some extent biassed by an in-felt need for cosmic companionship.' She had expressed similar thoughts in an earlier essay on Mars, following the publication of Percival Lowell's famous book in 1896 which firmly proclaimed the existence of intelligent life on that planet.[34] She found Lowell's book 'eminently readable' but 'as a contribution to science ... scarcely to be taken seriously'. The essay reviewed some less fantastic discussions of the same subject by Flammarion and Giovanni Schiaparelli. Though the emergence of organic life cannot be assumed as a matter of course, she wrote, 'the thought that millions of globes roll through space tenantless for all times revolts our sense of the rational in creation'. But science is limited to what it can observe; the rest is left to Infinite Wisdom. Agnes Clerke's views as well as Wallace's are included in M.J. Crowe's comprehensive history of the extra-terrestrial life debate up to 1900.[35]

Over and above these astronomical writings, Agnes Clerke continued her obligations to the *Edinburgh Review*. One must marvel at her industry as well as the breadth of her interest at this period in her life: revising her *History* and *The System of the Stars*, completing her work for the main *Dictionary of National Biography*, researching low-temperature physics for the Royal Institution and preparing her ambitious and unusual new book, *Problems in Astrophysics*. All of these projects were accomplished during the first three years of the new century.

Henry Hugh Peter Deasy

Meanwhile, another member of the family was achieving fame. In 1901, the Clerkes' first cousin, the 35-year-old Henry Hugh Peter Deasy, was awarded the Founder's Gold Medal of the Royal Geographical Society for his explorations in Tibet.

Henry was the younger son of Judge Rickard Deasy, Mrs Clerke's brother. Rickard Deasy had married fairly late in life and his children were thus considerably younger than their cousins. Of these, two sons survived. The elder, a cadet in the Kerry Militia of the British Army, died tragically in 1881 at the age of 19. A moving letter from the father to John Clerke recounting this sad news testifies to the close family bond that united the two families.[36] The Judge himself died in 1883.

Henry, the sole survivor at only 17 years of age, in due course joined the British Army, rising to the rank of Major in the 16th Queen's Lancers. At his Preparatory Boarding School, St Aloysius in Bournemouth, he had been awarded at the age of 14 a prize 'for general excellence'; it was Edward Whymper's *The Ascent of the Matterhorn*. This no doubt it was that generated his 'long wish to explore'. He sold the magnificent mansion at Carysfort which he had inherited from his father to finance an expedition to the mountains of Tibet. Fittingly, the new owners of Carysfort were the Sisters of Mercy in the Baggot Street Convent, who acquired it as a Teacher Training College for women. When home in 1893 on sick leave from India where his regiment was stationed, Henry took up the study of surveying under the guidance of a member of the Royal Geographical Society, and also of astronomy (though nowhere does Agnes mention his name to any of her correspondents). On voyages to and from India he practised taking astronomical sights at sea, and in India studied surveying under officials of the Trigonometrical Branch of the Indian Survey at Dehra Dun. While in Britain he tested his skill at determining longitudes by the method of observing moon-culminating stars at the Royal Observatory Greenwich, where following a series of observations over five nights he estimated his error at less than 1 second. His published Report includes an acknowledgement to Mr Christie, the Astronomer Royal.

Deasy assembled his equipment, which included a collapsible boat, loads of ammunition and sheep as beasts of burden, and set out, accompanied by two orderlies from the Gurkha Regiment and Indian helpers, from Shrinagar in 1896. After this first exploratory expedition he resigned his Army commission, and in March 1897, now entirely independent, commenced his second, longer, excursion. In the course of this he recorded the geographical latitude and longitude of 400 sites and the heights of some 100 peaks and passes. Botanical and zoological specimens were collected, and numerous photographs taken.[37] The enterprise, which lasted almost 3 years, was gruelling; when it was over, the intrepid explorer had to spend some time in hospital in Simla to regain enough strength to return to Britain. He reached London in December 1899 'so weak in health that I was scarcely able to crawl', as he wrote afterwards. However, it was all worthwhile. Henry presented a large number of botanical specimens to Kew Gardens and to the Natural History Museum, including some new plants that were given his name. His zoological specimens, which included a previously unknown rodent, were the subject of a paper published by the Zoological Society of London.

Henry was awarded the gold medal of the Royal Geographical Society in 1901. In the same year he published a fascinating record of his experiences in a book illustrated by his photographs.[38] He was in demand as a lecturer, one of his audiences in 1902 being Manchester Geographical Society;[39] the notes for his lecture on 'Exploring the Roof of the World', illustrated by lantern slides, still survive.[40] Henry's geographical exploits were warmly praised by a geologist reviewer in *Nature*.[41]

Henry Deasy now extended his enthusiasm to what he called 'motor mountaineering'. He became an expert motor car driver, recommending the use of the motor car for exploring the Alps, and performed some feats of driving on steep mountain roads. He eventually (in 1906) set up his own Motor Manufacturing Company in Coventry.[42]

It is not known whether Agnes or Ellen Clerke attended their cousin's illustrated lecture on 'Motor Mountaineering in the Alps', delivered to the Ladies' Automobile Club in London on 7 February 1905!

12 *Problems in Astrophysics*

The *Magnum Opus* 1903

Agnes Clerke's third and last major book was *Problems in Astrophysics*. This was the book that she had told Gill, as far back as September 1894,[1] was 'haunting' her, and which she felt 'driven to try'. It was to be very different from her other books, and very ambitious – not an account merely of past events and achievements, but of desiderata and of ideas for future research. It was to be, she believed, her *magnum opus*.

It would appear that Agnes Clerke began writing in earnest at the beginning of 1898, after her mother's death. Gill's reaction, on hearing that the book was in progress and to be finished in a year, was:

> 'What a woman!' and then 'what a foolish woman'. I meant by 'foolish woman' that you were risking not only health but opportunity by finishing such a work in so short a time; time to do what I expect you to do in that book in a year must either kill you, or fail to do you justice, or both.[2]

She explained the plan of the book, which would include besides the sun only stars and nebulae.

> It is not bound to be exhaustive. Planets involve very different considerations belonging at present mainly to the telescopic department . . . So I think of leaving them for another volume, should my powers last long enough to reach it.[3]

She had already drafted the chapters on the sun, fully realising however that they might have to be re-written in the light of observations of the total eclipse of that year, 1898, observed in India. Agnes Clerke was to see the results of various expeditions exhibited at a Royal

Society Soirée. During the summer of 1898 she worked on the book taking no holiday, apart from a few days which she and her sister spent with the McCleans in Tunbridge Wells. 'Absolute quiet I feel a necessity of life, almost. But I am recovering very slowly and must work as much as my powers admit of.'[4]

A year later (February 1900) Agnes Clerke was struggling with the book through a bout of influenza 'which has not helped the progress of my magnum opus; but I am regaining health though slowly';[5] after almost another year, with 'a lot of work on hands and not much vigour for getting it done', a rough draft was completed and revision was about to start.[6] She hoped to have the book finished by the end of the year, before tackling a fourth edition of the *History* which the publishers were calling for.[7] When the manuscript was finished she asked David Gill if she might dedicate the book to him. 'As you gave me the first idea and the incitement to write it, I should like to offer it to you publicly, if you do not object' – though it was not strictly true that Gill had suggested it. 'I have no idea how it will turn out. It was composed during a period of great depression and lassitude of which it cannot but show the traces. Still, I believe there will be nothing in it that you will dislike being associated with'. 'I'll be the proud man!', was Gill's response.[8] 'May the work which you have so long enhanced fully reward you – as I think it will.'

In the event, while Agnes Clerke arranged the illustrations for *Problems*, the fourth edition of her *History* came out (1902), a copy of which she sent to Gill on August 26 of that year.[9] The appearance of *Problems in Astrophysics* was thus delayed until 1903, its preface being dated December 1902. It was 'Dedicated by permission to Sir David Gill, K.C.B. whose suggestion and encouragement prompted its composition and animated its progress'.

Problems in Astrophysics

Agnes Clerke's *Problems in Astrophysics* was, in her own view though perhaps not in posterity's, her greatest accomplishment. She had spent many years on it in the face of illness and depression. At times she was only able to toil for half-hour periods, and feared she would not live to complete it.[10] Her dutiful motive in writing it was, one presumes, to place her encyclopaedic knowledge of astronomy, and her appraisal of it, at the service of working astronomers.

'The object of the present work is not so much to instruct as to suggest', were the opening words in the book's preface. She continued with a military metaphor:

> It represents a sort of reconnaissance, and embodies the information collected by scouts and skirmishers regarding practicable lines of advance and accessible points of attack, with a view to annexing for the realm of knowledge some further strips and corners from the territory of ignorance. Its inspiring motive, in short, is the desire for a rectification of the frontier in the interests of science.

In this respect *Problems in Astrophysics* differed from *The System of the Stars*, which was basically an assembly of established observational facts. *Problems*, by contrast, laid bare the unanswered questions. In preparing the earlier book – which she was now updating – Agnes Clerke had begged her eminent mentors at Lick and the Cape to supply her with material and to give advice which she gladly accepted: indeed, the text had been read, annotated and approved by David Gill, before being sent to the printers. The new book was entirely her own project, unseen in advance even by David Gill, in which she put forward her own views of what astronomers ought to be doing with their powerful instruments. 'The globe is studded with observatories, variously and admirably equipped. Yet innumerable objects in the sidereal heavens remain neglected, mainly through inadvertence to the extraordinary interest of the questions pending with respect to them', she wrote in her preface. The thought of under-used facilities had always bothered her. She admired the Lick astronomers for their efficiency: 'Your great telescope is certainly pre-eminent no less for activity than for size, and goes far to make up for the uselessness of many fine instruments in all parts of the world', she once wrote to Edward Holden[11]. The motto on the title page, from Francis Bacon's *Novum Organum*, expressed this cry for more observations: '*In solidis et veris aspiramus ad ultima et summa*'.[12]

The book was composed of two parts, the first, occupying one third of the whole, devoted to solar physics, and the second to the physics of the stars and nebulae. This division of subject matter was conspicuously different from the *System* which treated the sun as a star among many, not as a special object. Neither book dealt with planetary astronomy, which still belonged to 'the theoretical and descriptive departments of the elder celestial science'.

The section on the sun recorded the astounding progress made during the 1890s in solar observations. Most dramatic of these was the invention of the spectroheliograph in 1892, which revealed the detailed structure of the chromosphere and the motions of prominences beyond the sun's limb. Unfortunately, it was published too soon to include the advances of Hale's new spectroheliograph, operational from 1903. Henry A. Rowland's *Preliminary Table* of 20,000 solar wavelengths had been published and almost forty chemical elements, including the elusive oxygen, had been recognised in the sun's spectrum. Three total solar eclipses had been successfully observed; spectacular photographs of the corona and the 'flash spectrum' abounded. In her usual systematic manner, Agnes Clerke considered separately each of the layers reputedly making up the surface of the sun – photosphere, reversing layer, chromosphere, corona. There was also the 'dusky veil', between the photosphere and the reversing layer, allegedly responsible for limb-darkening.

Recent eclipse observations had highlighted a major dilemma with regard to this model. The flash spectrum, successfully photographed by several observers, notably Lockyer and John Evershed, proved not to be an emission line version of the sun's dark line (Fraunhofer) spectrum, as would be expected by Kirchhoff's law if it were emanating from that same layer. 'Here it was assumed that the missing rays were stopped, and here also it was assumed that the missing rays would be seen bright, could they be isolated', Agnes Clerke had earlier said in her *History*.[13] A striking pair of photographs reproduced in the book showed the spectra of the last sliver of the sun's photosphere before the instant of totality and the flash spectrum immediately following. Some, but far from all, of the lines were the same. Agnes Clerke tried to find explanations for this puzzle but had to say that 'the difficulties were neither few nor trivial'. There was also the long-recognised problem of the spectra of the chromosphere (placed above the reversing layer like 'a gaseous ocean' upon which the prominences develop 'like mounting waves and tossed spray') and of the corona. Helium, a constituent of the chromosphere where it was first discovered, and 'coronium', a hypothetical element responsible for a conspicuous green line in the spectrum of the corona, were both entirely missing from the sun's Fraunhofer spectrum. Non-eclipse observations of the spectra of eruptive prominences by the leading solar physicists of the day – Hale, Deslandres and Evershed – showed emission lines associated in the laboratory with high temperatures. 'This has been taken [by Lockyer] to imply that the chromospheric is essentially a high temperature spectrum', she wrote – an explanation to be rejected out of hand: it was taken as certain that temperature dropped with height above the solar surface. (The new physics of a later generation would show that the temperature of the sun's atmosphere does indeed increase with height and that the unfamiliar emission lines in the spectra of the chromosphere and corona are due to highly ionised gases.)

Lockyer, as history would show, had made a very important contribution to spectroscopy with his dissociation theory of the elements, enunciated in 1878. He had put a great deal of effort over many years into investigating the spectra of chemical elements under various laboratory conditions and had demonstrated that when the temperature is raised spectra exhibit 'enhanced lines', different from those emitted under cooler circumstances. This could be done by using an electric arc and the much hotter electric spark to produce the spectra. He concluded that the atom was not the ultimate particle, but broke down or became 'dissociated' at high temperatures. What Lockyer was observing at the higher temperature was gas in its ionised state; but this explanation was not forthcoming until the age of atomic physics. In fact, Lockyer's insightful theory prefigured Saha's theory of ionisation of 1920.

In the first edition of *System of the Stars* Agnes Clerke explained Lockyer's theory very well by citing the strong lines, H and K, of calcium in the solar spectrum.

The pair form the most strongly marked feature of the spectrum of calcium when raised to the highest pitch of incandescence: and there is much to be said in favour of Mr Lockyer's view that they emanate, not from calcium in its entirety, but from some of its subtler ingredients. There is no doubt at any rate that they are what is called 'high temperature lines'; the light of ordinarily glowing calcium does not contain them.[14]

The H and K lines are due, as would eventually transpire, to ionised calcium. In *Problems in Astrophysics*, Agnes Clerke was less sure; her friends the Hugginses and the physicist James Dewar and his spectroscopist colleague George D. Liveing were among those who denied the temperature effect claimed by Lockyer. The Hugginses put it down to the effect of low pressure in hotter stars, which was indeed a factor.

A historically important chapter of the book was devoted to the problem of the temperature of the solar photosphere, one of the first such in the literature. She described the best available measurements of the sun's total radiation from which the realistic value for the sun's effective temperature (6590 degrees) had been obtained using Stefan's Law by W.E.Wilson of Daramona – though she was not entirely convinced of the authenticity of the calculation; she opined that Stefan's Law was 'not unconditionally to be trusted'. As to Wien's Law, which stated that temperature was inversely proportional to wavelength, it was 'suspiciously simple, one of those formulae which cannot be trusted far out of sight'. Other chapters dealt with sunspots, solar rotation and the sunspot cycle.

Progress in stellar astrophysics, particularly spectroscopy, had been spectacular while the book was in the making. At Harvard, the spectral classification programme was in full swing and variable stars were being discovered on photographs as a matter of routine. Other observatories, notably Lick and Potsdam, were successfully observing line of sight velocities in stars and adding to the growing list of spectroscopic binaries.

Spectral classification had begun with Secchi's four types (1866), forming what Agnes Clerke called 'irremovable landmarks'. Other refinements followed. In the *Draper Memorial Catalogue* of 1890, the

Secchi types were subdivided according to Williamina Fleming's system and labelled by letters of the alphabet. The most recent classification system, also at Harvard, was due to Antonia Maury, who divided the spectra into 20 divisions and introduced subdivisions according to the width of the spectrum lines. Agnes Clerke chose to pass over all of these post-Secchi systems. 'There is a danger of stellar classification degenerating into a maze of provisional distinctions. The best remedy is to fix attention on the summit-ranges of the landscape; when they are clearly imprinted on the mind, mastery of detail can be safely and readily acquired.' She proposed a system of her own, of just eight recognisable compartments, accommodating practically all the stars. A chapter each was then devoted to a description of these types.

The question of spectral classification was bound up with the problem of the evolution of stars. In a general way it was thought that all stars had a similar life history. It was agreed that nebula-embedded stars, emission-line stars and the hot helium stars were the most recently formed. From there, the stars were thought to evolve from one spectral class to another until finally cooling off and disappearing into the realms of the unobservable. Lockyer's meteoritic theory was not mentioned. By the end of the century the initial enthusiasm for it had died away.[15] However, as Agnes Clerke laid out in a chapter on stellar evolution, the life story of individual stars on the evolutionary track could vary greatly. She discussed the theoretical work of Homer Lane on the contraction of gas spheres, which predicted that such contraction was not necessarily accompanied by a fall in temperature. Stars of different masses could lose their energy at different rates. Members of binary systems might have different life expectancies. The picture was confusing, but then the purpose of the book was to expose such problems.

Later chapters discussed double stars, spectroscopic binaries, eclipsing binaries and various kinds of variables. A vast amount of information was concentrated into these chapters. The full history as well as every known observational detail of dozens of individual stars were recorded, pointing out what still needed to be done.

A seemingly insoluble puzzle was that of the short-period variable stars, in particular those of which Delta Cephei was the prototype, later known generically as cepheids. These stars had regularly varying light curves. They showed one spectrum only, with a radial velocity variation of the same period. Two stars, it was assumed, were involved, one bright, the other dark. The minimum of light could not be explained, however, in terms of the eclipse of the bright member by the dark one: when the light was a minimum the velocity was a maximum, not a minimum as would be the case in an eclipsing system. Astronomers, however, were determined to fit a square peg into a round hole by insisting that the variations in velocity and light could somehow be explained by two or more stars. Agnes Clerke had written an account in the *Observatory* of an apparent solution by the Russian astronomer and radial velocity expert A.A. Belopolsky.[16] She agonised over the matter in her new book but could only conclude that they are binaries, but not eclipsing pairs. The true explanation for these and also for the Mira types – single pulsating stars – did not come until the 1930s.[17]

With very little understanding of the difficulties and delays of observational astronomy she complained of objects being 'pigeon-holed for future reference . . . and in their pigeon holes they have been allowed to lie'. She took the opportunity of listing the variable stars which she had asked Pickering, in vain, five years earlier, to test at Harvard for eclipsing binary properties. 'When the spectra of [these stars] have been examined, and their light changes tabulated and collated, we shall be in a better position to interpret those manifest in the [Beta] Lyra variable.'[18] But the book came out, and no-one leaped into action. 'Harvard College has to bear the brunt of my wrath', she complained to Gill.[19] Needless to say, her 'wrath' was for Gill's ears only: Agnes Clerke was the mildest of people, unfailingly polite to all her correspondents. The stars in question proved in the course of time to be either cepheids or eclipsing binaries.

In the midst of all the perplexity, the book had a modern tone in that wider topics in physics were introduced, even if not fully understood. Ideas aired included the kinetic theory of gases and Clerk Maxwell's theory of heat, Herzian waves (a continuation of the infrared which had been searched for but not found in the sun, because, she remarked, they were probably arrested by the earth's atmosphere) and the Zeeman effect. She drew attention to various 'theories' which in the state of physics at the time were hardly more than opinions: Schaeberle's theory of the solar corona, which saw it as caused by 'streams of matter ejected from the lower latitudes of the sun', and William Huggins' 'electrical theory' which saw coronal streamers as 'analogous to comets' tails, issuing forth under the influence of a repulsive force'. The 'triple origin' of coronal light was recognised – 'continuous reflected light is mixed up in it with continuous original light, and these again with bright emission lines', with the proportions of these lines varying.

The second part of the book, 'Problems in sidereal physics', dealt with star clusters and nebulae, including 'white nebulae' (galaxies), the final chapter being devoted to 'The physics of the Milky Way' and the distribution in space of the various types of object. Agnes Clerke retained her view of the universe with its evolving constituents as entirely within the confines of our own galaxy. She ends the book with a quotation from Kant: '*Die Schöpfung ist niemals vollendet. Sie hat zwar einmal aufgegangen, aber sie wird niemals aufhören'* (Creation is never completed. It started at one time, but will never cease.)[20] Creation, in her vocabulary, meant the continuous process of star formation and evolution.

Problems in Astrophysics, a bulky work of almost 600 pages with numerous diagrams, was illustrated by 31 photographs from various sources, not all of them from the famous observatories. The frontispiece, in tribute to her best friend, was David Gill's photograph of Eta Carinae crisscrossed with a measuring grid. She had intended to include Keeler's photographs of nebulae taken with the Crossley reflector at Lick which he had given her in 1899 and which had left her gasping in admiration, but unfortunately for administrative reasons Keeler's successor Campbell was not allowed give permission.[21] She was 'thrown back upon any that could be hastily collected to supply their place'.[22]

The substitutions were three plates of nebulae photographed by her friend W.E. Wilson with his 20-inch Grubb reflector at Daramona. Those chosen, though beautiful in themselves, included only one relevant spiral nebula (M51, the whirlpool) (Figure 12.1) to replace the three

Figure 12.1 The Whirlpool nebula, by W.E. Wilson.

Keeler spirals of the original plan (M51, M101 the pinwheel, and NGC 7479 in Pegasus). Other photographs came from Isaac Roberts (Figure 12.2), Barnard and Pickering.

The chief interest of *Problems in Astrophysics* for the modern reader is as an illustration of the confused state of theoretical astrophysics before the dawn of the new physics.

The book welcomed

David Gill, on receiving his copy of the book, had no doubts about its excellence. 'So happy, so strong, so useful a book. . . . I do not believe there is a man living who knew beforehand all the facts that you have brought together, and brought together so well in their proper places.'[23]

W.W. Campbell got in touch with Hale, editor of *Astrophysical Journal*, offering to review it. Hale was delighted. 'I also have been greatly interested in Miss Clerke's last book . . . All sorts of interesting questions are raised in this book, and it ought to prove very stimulat-

Figure 12.2 The globular
cluster M11, by I. Roberts.

ing.'[24] 'Whatever you say about it will be of special value to the public
and instruction to me', Agnes Clerke told Campbell on hearing that he
was to be the reviewer.[25] He did not disappoint her. He gave it his enthusiastic backing in a ten-page report, Hale having told him to make it as
long as he liked. Referring to the speed of development in astrophysics
in the previous 15 years and the need for a critical survey of the field, he
declared that 'the appearance of Miss Clerke's book could scarcely have
been better timed', and that her stated purpose ('not so much to instruct
as to suggest') had been fulfilled. 'By way of suggestions for future lines
of research this book is the richest one known to me'. Her account of
progress in solar physics he described as 'masterful'. Campbell was also
impressed by the chapters on variable stars. He agreed with her that

knowledge of their spectra was remarkably slight. 'Miss Clerke empha-
sises the need for more work. There is in my opinion no richer field
awaiting systematic cultivation.' He found only a few minor faults, and
regretted, with justification, that the illustrations of clusters and
nebulae were not the most modern available.

He summed up his long and flattering review by saying that the
book was 'ideally planned, the style lucid, attractive and logical'. 'To
me its most remarkable feature is the wonderfully correct estimate of
the relative values of observations, by an author with little or no experience in making observations.'

At home, the *Observatory* gave the new book 'a respectful and
hearty welcome' with a review by its editor Henry P. Hollis.[26]

Nature's comments

Nature, by contrast, took a very different line. A detailed account of the
book entitled 'The spectroscope in astronomy'[27] was signed by R.A.
Gregory (later Sir Richard Gregory), who had in all likelihood been the
author of the cool notice of the third edition of her *History* in the same
journal 10 years before. Gregory, though trained in spectroscopy under
Lockyer, was no longer a practising scientist, being fully occupied with
his writing and the management of *Nature*. Having outlined the contents
of the book – 'related in the exuberant style with which all readers of
astronomical literature are familiar' (a doubtful tribute!) – he proceeded
to pour scorn on the author and to belittle her knowledge of astrophysics.

> As is the case with every branch of science in its youth, questions arise
> much faster than they can be answered, and it requires a fine critical
> faculty to separate results of transient value from those of significance
> to scientific progress. The historian has to decide what things matter
> and what may be neglected from their influence upon development;
> and success is achieved when this power of discernment is combined
> with insight which enables the relationship to be seen between cause
> and consequence. With the best will in the world to give Miss Clerke
> credit for her work, we must confess that it is not altogether
> satisfactory.

Then followed a list of her faults: results are 'of little value or immature', 'she has not understood the real nature of some of the material collected' and 'she passes judgment and gives advice on matters which can only be rightly understood by investigators actively engaged in spectroscopic work'. While, according to him, 'a man who has had a scientific training can quickly grasp the essential points of progress', Agnes Clerke, with her lack of 'personal and practical acquaintance with the subject . . . should remember that "Passengers are respectfully requested not to speak to the man at the wheel"'. This last was a reference to Agnes Clerke's complaint that 'innumerable objects in the sidereal heavens remain neglected', as if instructing astronomers in their duties.

It could well be argued that Gregory had a valid point here, though he might have expressed it less crudely. Agnes Clerke may indeed have had an unrealistically confident opinion of her own judgement on practical and policy matters. However, the book did not strike her distinguished friends and 'men at the wheel' as at all arrogant; on the contrary, they regarded it as a useful review of contemporary fields of astrophysics, packed as it was with references.

Gregory completed his review with an unchivalrous gibe. 'A cynic has said that it is a characteristic of women to make rash assertions, and in the absence of contradiction to accept them as true. Miss Clerke is apparently not free from this weakness of her sex.' This remark was backed by particular reference to his view that Lockyer's meteoritic hypothesis was not adopted as an explanation for dark prestellar matter. That was true; but she had not overlooked his work of more recent vintage. Indeed, she and Lockyer never quarrelled.

Agnes Clerke was not over-concerned about criticism. 'Personally, I am very *in*sensitive about reviews' she told Gill on an earlier occasion, adding that the Hugginses, who were often attacked by the Lockyer camp, were 'unluckily quite the reverse'.[28]

In the literary world, *Problems in Astrophysics* was the subject of an erudite article entitled 'The new astronomy' in the *Edinburgh Review*.[29] It was also reviewed in *The Academy*,[30] a weekly review of literature, science and art. The writer was the poet Francis Thompson, a lover of astronomy and the author of a poem, 'A dead astronomer', written in memory of Father Stephen Perry who had died of fever on his way home from observing the eclipse of 1889.[31] He declared: 'Miss Clerke's writing is clear in the extreme, and would almost render her book popular were popularity possible with subject-matter so recondite'.

Member of the Royal Astronomical Society

Agnes Clerke's new book swiftly brought its reward. According to one newspaper commentator,[32] '*Problems in Astrophysics* was of such great scientific value that the Astronomical Society could no longer ignore her claims to public recognition by them'. At the May meeting of the society, by the council's decision, Agnes Clerke and Lady Huggins were made honorary Members of the Royal Astronomical Society. The status of honorary membership had been previously given to only three women – Caroline Herschel and Mary Somerville in 1835 on the strength of their scientific achievements, and Anne Sheepshanks in 1862 as a benefactress of astronomy. In announcing the election the President of the Society James Glaisher declared:

> It is a pleasure to think that there is a considerable resemblance between the claims of these ladies and those of our original honorary members. Lady Huggins has been associated with the work of her husband as Miss Caroline Herschel was associated with the work of her brother. The work of Miss Agnes Clerke is similar to that of Mrs Somerville, lying in the domain of scientific writing, and, I may say, with reference to her last work, it is not merely an astronomical history, but a work of actual constructive thinking in our science.[33]

The recognition was long overdue; Agnes Clerke was over 60 years of age and Lady Huggins was 55. William Huggins, recently decorated with the Order of Merit and President of the Royal Society, was, with his wife, soon to retire from active research. Whatever their inner thoughts on the matter, however, the two new honorary members were pleased thereafter to append the letters Hon. Mem. R.A.S. to their names.

13 Women in astronomy in Britain in Agnes Clerke's time

The world scene

The election of Agnes Clerke and Margaret Huggins to the Royal Astronomical Society, almost 70 years after their only two predecessors were similarly honoured, would appear to indicate that women played a very small role in astronomy in nineteenth century Britain. That indeed was the case. An international register compiled early in the new century listed all observatories, astronomers and their instruments operating world-wide at that time.[1] It showed that those engaged in astronomy or allied fields (geophysics, navigation, meteorology) in institutions throughout the entire world numbered fewer than a thousand. In Britain, the number of people in permanent paid posts in astronomy barely exceeded 100, of whom half were computers.

A conspicuous aspect of these statistics relates to the position of women. The United States, in contrast to all other countries, employed women on the staff of their observatories for astronomical research. They were part of the accepted scheme of things. Nineteen names of members of the well-established women's team of spectroscopists and photometrists at Harvard are recorded. Each of the four all-women colleges in the United States also had two or three female astronomers on its academic staff, while a few further women were employed as computers at other institutions. The grand total was fifty.

The history of the Harvard scheme, which started in 1875, has been told many times. Though the women were not highly ranked nor well paid, they were doing cutting-edge research and constantly making discoveries in the fast-growing field of astrophysics.[2]

The situation in the rest of the world was very different. Women, where employed, were anonymous workers on the international *Carte du Ciel* project (in which the United States did not participate) initiated in Paris in 1887 (see Chapter 6). The great aim was to survey the entire sky photographically, recording stars down to magnitude 14 and providing a catalogue (*The Astrographic Catalogue*) of all stars down to magnitude 11. The international community of astronomers had agreed to share this huge task between eighteen observatories throughout the world, each observatory being allotted an agreed zone of the sky and each observatory to be responsible for measuring its own photographs. Two observatories only in Britain – Greenwich and Oxford – participated.

The photographs on glass plates ran into thousands and the star images into millions. A common method of measurement and reduction was adopted whereby positions of star images on a plate could be converted to celestial spherical co-ordinates. The task required large numbers of measurers and calculators working under strict obedience to laid-down procedures. The employment provided was of a routine kind, carried out at most observatories by teams of up to ten women.

Apart from the *Carte du Ciel* teams, there were only three or four women listed as working at observatories in continental Europe, apparently voluntary helpers such as the wife of Max Wolf of Heidelberg, doing searches on photographic plates. Outside the United States, therefore, the opportunities for women to take part in genuine astronomical research was small or non-existent.

Confirmation of this general picture is provided by the indices of Agnes Clerke's well-researched major books – the *History*, *The System of the Stars* and *Problems in Astrophysics*. The pioneering Harvard women – Mrs Fleming, Miss Maury and Miss Cannon – occupy significant places in the books, as does Lady Huggins (in joint references with her husband). Apart from these, female names, all but one of them British,[3] are few.

Deservedly present are the solar observer Elizabeth Brown, one of whose fine sunspot drawings is reproduced in *Problems in Astrophysics*, and Annie Maunder, whose unusual coronal photograph is shown in the same volume. Violet Common (sister of the telescope-maker A.A. Common) is mentioned for the drawings she made of the corona of 1889 from a photograph by the late Stephen Perry, and

Gertrude Bacon, amateur astronomer and balloonist, for her photograph of the corona at the 1900 eclipse. A minor reference goes to Miss Airy, daughter of Sir George Airy, for her acuity of vision is being able to see twelve members of the Pleiades with the naked eye. Agnes Clerke, therefore, was at pains to place the contributions to astronomy of British women, however small, on record.

The Grand Amateurs

The inclusion or omission of a name in a historical account or on a register does not necessarily define a line of demarcation between the productive and the futile among astronomical labourers. Essential work behind the scenes of often talented assistants goes unrecorded.[4] Nowhere is this more likely than in the case of the contribution of female family members. Caroline Herschel and Margaret Huggins were exceptions, being acknowledged partners of epoch making pioneers (though Caroline receives only a footnote in Agnes Clerke's *History*).

The Victorian 'Grand Amateurs'[5] – serious astronomers of independent means and possessors of large instruments of whom the Hugginses were the last – were principally responsible for the advance of astronomical knowledge in Britain for much of the nineteenth century. Some are known to have been helped by their cultivated wives and daughters. The wife of Admiral W.H. Smyth, successful and inspiring amateur, and compiler of the famous Bedford double star catalogue, the *Cycle of Celestial Objects* (1844), was one of them. Maria Mitchell, in the course of her Grand Tour of Europe in 1857, visited the Smyths at the observatory of their wealthy neighbour John Lee at Hartwell House. She described their activities: 'In the absence of the Lees, he [Smyth] has a private key, with which he admits himself and Mrs Smyth. They make the observations (Mrs Smyth is a very clever astronomer), sleep in a room called 'the Admiral's Room', find breakfast prepared for them in the morning, and return to their own house when they choose'.[6] Mrs Smyth, a talented artist, provided some of the technical illustrations for her husband's book on Lee's observatory, *Aedes Hartwelliana*,[7] while the Smyth social circle included all the leading astronomers, professional and amateur, of the day.[8]

Other family helpers were the daughters of William Lassell, whose 24-inch telescope was donated to the Royal Observatory and figured in the offer of a computership there to Agnes Clerke. It may be not without significance that their mother was the daughter of a navigation and mathematics teacher.[9] 'The Misses Lassell, four in number, seem to be very accomplished', reported Maria Mitchell. 'They take photographs of each other, make their own picture frames, and work in the same workshop with their father.' She was pleased to learn that they also had observed 'her' comet.

The expedition wives

Expeditions to foreign lands were an important feature of astronomy throughout the nineteenth century. Elaborate efforts and considerable expense went into observing phenomena which could not be observed at home – such as the transits of Venus and total eclipses of the sun. Agnes Clerke opens her chapter 'Recent solar eclipses' in her *History* with the total eclipse of July 1860 observed in Spain, the first at which the solar corona and prominences were successfully photographed. The British expedition included the wife and daughter of the Astronomer Royal, G.B. Airy.[10] At the total eclipse of 1870 at Cadiz 'a lady' made a set of ten drawings of the corona, which accompanied Lord Lindsay's report to the Royal Astronomical Society of his expedition's observations. Norman Lockyer's wife, Winifred, accompanied her husband and his party on an expedition to observe the same eclipse from Sicily. They had a double dose of bad luck. The ship carrying them and their equipment struck a submerged rock soon after embarking from Naples, and though they escaped and were able to set up their base at Catania in time, the weather was against them.[11]

Far more strenuous for the wives concerned, and of more import, were two foreign expeditions for unique purposes. The first was Charles Piazzi Smyth's 6-month expedition to the island of Tenerife in 1856 to test the advantages of making astronomical observations from

a high mountain site.[12] Piazzi Smyth, W.H. Smyth's son and
Astronomer Royal for Scotland, was accompanied on his mission by
one helper only, his newly married wife Jessie.

Jessie was intelligent and well educated, brought up in affluent
surroundings on a country estate in Aberdeenshire. Among her interests was geology, which she studied in Edinburgh with the well-known
tutor Alexander Rose, whose two-year course included field trips to
local geological formations.[13] Afterwards, she made extensive geological tours not only in Scotland, England and Ireland, but also in
Switzerland and Italy. She was over forty when she met and married
Piazzi Smyth.

The Smyths set up their observing station and their living quarters under canvas near the Peak of Tenerife. Piazzi Smyth's extant
photographs show Jessie in charge of her telescope, though the wife's
name was not recorded in the official Report. That Report, which
earned Piazzi Smyth his Fellowship of the Royal Society, was to be
illustrated with some of his own photographs. This being refused on the
grounds of cost, Jessie, undismayed, ran off the required 700 photographic prints in her Edinburgh kitchen. These were incorporated in
the Report, each one inscribed 'CPS Phot. JPS Pr'.[14] The prints, now
well over a century old, are still in perfect condition.

Piazzi Smyth, unfortunately, was never able to persuade the
Government to finance a British mountain station. His later researches
in laboratory and solar spectroscopy were carried out at his own
expense, at home and on trips to sunny locations abroad, his wife
always by his side. He was to be congratulated, wrote one visitor, on
having 'a wife that can sympathise and work with you in your
researches' while another referred to 'the most intelligent and agree-
able assistant in the person of Madame Piazzi Smyth'. During a stay in
Sicily in 1872 the astronomers at the observatory of Palermo were
amazed to find Jessie up at night observing with her husband.[15]

The second expedition wife, also from Aberdeenshire, was Agnes
Clerke's future dear friend Isabella, the wife of David Gill. Mrs Gill's
experience as an astronomer's assistant, when as a young woman of
twenty-eight she accompanied her husband on their expedition to
Ascension Island in 1879, was short, but of immeasurable importance.

Though she was not required to make actual astronomical observations, Gill would not have succeeded without her. She trudged the island by night searching for and finding the best site for their camp, kept watch for breaks in the cloud, copied notes, sent reports back to London, and generally made herself useful. 'A considerable part of the success of the expedition was due to the unfatigued assistance [of Mrs Gill]', said Arthur Auwers.[16] On their return to London, Mrs Gill made her own special contribution, a delightful book called *Six Months on Ascension*,[17] the first published account of the geography and scenery of that remote island.

Unlike Piazzi Smyth's wife, Mrs Gill forsook active astronomy after this one experience. Nevertheless, like other astronomical wives who lived in the semi-monastic conditions of a residential observatory, this 'highly gifted wife' (Kapteyn's words[18]) could not fail to be well-informed about her husband's work. Surviving correspondence between other astronomers' wives confirms how closely such women were involved in the scientific culture of the Victorian age.[19]

The lady computers at Greenwich 1890–95

In 1890 the Royal Observatory took the unprecedented step of employing women on its staff for scientific duties (Chapter 6). Despite the designation 'computer' the women were not condemned to a treadmill but were assigned to work with individual assistants. Hours of work were 9.00–1.00 every weekday and 2.00–4.30 on three afternoons a week, plus observing three nights a week for up to 3 or 4 hours.[20] Two of the women appointed, Alice Everett and Annie Russell, were determined career women of the new generation.[21] Alice Everett was assigned to the photographic department where the *Carte du Ciel* programme was about to commence with the installation of the astrographic refractor. Here she was given responsible and interesting work, new to the observatory. Annie Russell joined the solar department, helping E.W. Maunder with the famous Greenwich photoheliographic programme. Outside their formal duties the women were permitted and apparently encouraged by their colleagues to use other instruments at the observatory and to report their observations to the British Astronomical Association.

Efforts to keep the remaining lady computer posts at Greenwich filled ran into difficulties. The hours were unsocial and the pay was low. However, this does not appear to have been the root cause of the demise of the scheme.[22] The experiment had a time limit of five years, and was due to terminate in 1895. The women became distinctly 'supernumerary'.[23]

It was not the fault of their male colleagues that permanent posts for women at Greenwich had not materialised. The young women received every encouragement from their Greenwich superiors and had the support of the astronomical establishment. The supernumerary computerships, being temporary, were the only appointments which could be made without breaching the rules of the Civil Service, but being so, they could not be made permanent. It was all the more disappointing for the two young women that their hopes to become members of the Royal Astronomical Society had been thwarted (by, it would seem, non-academic fellows of the society).

Alice Everett moved to the Astrophysical Observatory in Potsdam in Germany and began work there on 1 October 1895. Annie Russell had a personal solution: she married her senior colleague Walter Maunder, having resigned from her post at the end of October.

The Cambridge physicists

The sympathetic attitude of the Greenwich astronomers towards their women colleagues also existed among physicists. There was no prescribed route into research or academic posts in experimental physics for women, yet a number of talented women worked unobtrusively in one form or another at the Cavendish laboratory at Cambridge in the last decades of the nineteenth century.[24] 'The men at the Cavendish did not ridicule their presence but gave directions, suggestions, and occasionally marriage proposals', concluded Paula Gould from a study of early Cambridge women physicists. It was a picture of collaboration not confrontation. Some of the women were daughters or sisters of academics; all came from strongly supportive families. This was also the case with both Everett and Russell. Alice Everett's father, professor of Physics at Queen's University Belfast, had long associations with the Royal Observatory. Annie Russell resided with her brother and sister, both medical students in London, in a house close to the observatory, undoubtedly provided for them by their parents.

Everett and Russell are examples of university-educated women for whom the opportunity to work in an active scientific environment far outweighed the lowly rank of the posts on offer. That such opportunities were rare is evident from a study of the register of Girton College, which records the subsequent careers of its former students.[25] Of just over 200 women who took Cambridge tripos examinations in the decade 1880–89 (who include Everett and Russell) about half (96) took either mathematics or natural sciences. Of these, sixty-two became school teachers, but only twelve had posts which were in any way associated with the academic world, of whom the Greenwich women were two. Six of the women went on to qualify as medical doctors, part of a more general involvement of educated women in philanthropy and social causes.

Subsequent careers of the lady computers [26]

When Alice Everett went to Potsdam in 1895 she was the first woman to be officially employed in an observatory in Germany.[27] Her post, on a staff of ten, was that of scientific assistant working on the *Carte du Ciel*, the same work that she had been doing at Greenwich. It was unfortunately no more permanent that the one she had left, being a 3-year replacement for someone absent on military service, but she worked hard and efficiently.[28] On finishing at Potsdam, she found employment for one year at Vassar College, USA, the women's college where Mary Whitney, successor to Maria Mitchell, was professor of Astronomy. This small institution had at that time only one other member of staff, and Alice Everett no doubt was pleased to have the opportunity of working there, even temporarily. Her year was fruitful and resulted in two papers jointly written with Mary Whitney on observations of minor planets and a comet in the *Astrophysical Journal*.[29] In 1899 she applied for a position to James Keeler, Director of the Lick Observatory. Keeler was anxious to employ her, 'a lady of distinction in astronomical science, admirably qualified to aid us in a most important part of our work – the measurement of our photographs of star spectra' but was unable to find the necessary funds. In 1900, at the age of thirty-five, she was back in England again, without a job. She had to abandon her heroic attempt to remain in astronomy and moved into the field of optics, where she eventually succeeded in making a satisfactory professional career.

The later career of her colleague Annie Maunder took a different turn. It was bound up closely with that of her husband. E.W. Maunder was a widower approaching 45 years of age with a family of five children, the youngest only seven years of age, when he married Annie, aged twenty-seven. Walter and Annie had no children of their own.

Maunder had been for some time disaffected at Greenwich,[30] understandably so, perhaps, as he was now the oldest assistant on the staff in terms of both age and years of service and likely to be resentful of the Astronomer Royal's bevy of brilliant young mathematicians recruited in the previous few years. To compensate, Maunder created his own empire in the British Astronomical Association. To him more than to anyone, with his 'silver tongue' and his tireless pen, the Association owed its tremendous success. Its first organised total eclipse expedition, to Norway in 1896, was largely his doing. It was on the Association's next eclipse of January 1898 in India, that Annie, using equipment designed by herself and paid for from a research grant from Girton College, obtained the photograph of the coronal streamer singled out for special praise by Agnes Clerke[31] (Figure 13.1). Later writers appear to have overlooked this remarkable observation.

It was not to Annie's advantage that her husband tended to publish accounts of his and her work under his own name in popular magazines and books rather than in the scientific journals. Had she been a Fellow of the Royal Astronomical Society she might have published her work on her own account.[32] The contributions to solar physics principally associated with the name of Maunder are twofold: the analysis of the cyclical variation in sunspot latitudes (the 'butterfly'

Figure 13.1 The solar corona photographed by Annie Maunder in 1898 with its long south-west rod-like extension. The original negative showed this reaching almost to the bottom right edge of the photograph.

diagram) based on almost 30 years' of continuous observations of sunspot positions on the solar disk, and the discovery of a 27-day periodicity in terrestrial magnetic activity associated with the synodic period of the sun's rotation, both published in 1904. Some of the papers in the series were published in the joint names of E.W. and A.S.D. Maunder. The basic work consisted in the analysis of accumulated solar and magnetic data, which of course Annie could perform at home. It is impossible to tell how far Annie (like Lady Huggins) was an instigator of the programme. She, working alone, used the same material to compile a massive catalogue of sunspot groups over a 30-year interval, which languished as an appendix to the official Greenwich Observations[33] – another instance of results that ought to have been published in a more visible place.

Other British women in astronomy

Though Alice Everett was involved in setting up the astrograph at Greenwich and in the early measurements, Greenwich did not have a designated *Carte du Ciel* team. Only one woman in the whole of Britain, the industrious Ethel Bellamy at Oxford, was employed on the project from the beginning. She was the niece of the observatory's chief assistant who joined the staff in 1899 at the age of seventeen and worked there for 50 years. Later, in 1909, a *Carte du Ciel* female team of four was formed at Edinburgh. Edinburgh, not an original participant, came to the aid of Perth Observatory, Australia, which had found the task too much for its means. The use of women who were prepared to accept relatively low salaries was 'intended to economise on the time of the scientific staff'. The Edinburgh women were paid £30 a year and worked in the forenoons only.[34]

At Cambridge Observatory Anne Walker worked for 24 years (1879–1903) from the age of seventeen, helping the observatory's assistant astronomer, on a salary, generous compared with that earned by the Greenwich and Edinburgh women, of £100 a year.[35] A second post at Cambridge was filled by a succession of four women computers between 1876 and 1904.

The rise of the British woman amateur

The foundation of the British Astronomical Association gave scope to women devotees of astronomy (see Chapter 8). As members of the Association they took a keen interest in observing variable stars and planets with their small instruments, reported meteors and aurorae, listened to lectures and published their observations in the Association's journal.[36] Astrophysics for the most part passed them by. Spectroscopes, photometers and photographic apparatus were beyond the means of the ordinary amateur. Their activities, therefore, are not included in Agnes Clerke's accounts of the progress of astrophysics (though they surely would have been, had she lived to write her promised volume on the planets). Visual solar system astronomy, on the other hand, all but abandoned by the professionals, flourished among the amateurs, something for which modern historians are now grateful. The Association's eclipse expeditions, when men and women could combine a scientific experience with travel to exotic locations, were especially enjoyable – as these events still are – and could occasionally provide significant results.[37]

A trawl of 'women astronomers' in the broadest sense (meaning women involved in astronomy in any capacity) between 1890 and 1930 throws up an estimated 107 names for the decade 1890–1900 and 150 for the following one.[38] Some of the enthusiastic women amateurs, it has been suggested, deserved the opportunity of making careers in astronomy. Annie Maunder, with her impressive record as an observer and as an active organiser in the British Astronomical Association, is cited as such an 'obligatory amateur'. 'One fact precluded her from becoming a professional: her gender', concludes one historian.[39] Certainly, had she been a man, Mrs Maunder would have been able to retain her job as a computer at the Royal Observatory in Greenwich on a salary of £4 a month, but this would hardly have made her a full 'professional'. Assistantships at Greenwich, of which there were only ten altogether, were plum appointments, reserved for top Cambridge mathematicians. Highly-qualified women mathematicians from Girton College, Cambridge – Hertha Ayrton, medallist of the Royal Society and wife of the engineer W.E. Ayrton (see Chapter 7), and Grace Chisholm Young, a PhD in mathematics from the University of Göttingen (the first woman to graduate from that university) and wife of the Cambridge mathematician Henry Young – never held professional appointments. Even today, Annie's 'senior optime' (second class honours) would not guarantee a place in astrophysical research. There were many 'obligatory amateurs' among men, too – William Shackleton, the spectroscopist who acquitted himself so well at the eclipse in Norway but was unable to find an astronomical post afterwards, comes to mind.

The lack of salaried employment for women was unavoidable (and for the majority of 'leisured amateur'[40] women, not a serious calamity). What could have been avoided, but sadly was not, was the exclusion of women for too long from intellectual parity with their male colleagues in the Royal Astronomical Society.

Women writers

Writing was always a career opportunity for pre-emancipation women. Mary Ward, first cousin of the third Earl of Rosse, constructor of the Leviathan at Birr, was the author of *The Telescope* (1858), an excellent book, reprinted several times, intended for the ordinary amateur observer and beautifully illustrated by her own drawings.[41] She possessed nothing more than a good-quality two-inch Dollond telescope, recommended to her by the Earl of Rosse, which she showed was capable of revealing a surprising number of the wonders of the heavens.

Another literary genre was translation. Middle-class women in the nineteenth century, who were traditionally taught European languages, could put their talent to use by translating foreign scientific books. Mrs Lockyer's timely translations from the French when few scientific books were being written in English included *The Forces of Nature*, by Amédée Guillemin[42] of Paris, author of the well-known *The Heavens (Le Ciel)*.

Elementary books on science for children were greatly in demand in the Victorian era in families where children were educated in the home. Here, women amateurs could play their part. In Agnes Clerke's generation, Agnes Giberne, who acquired her literary taste from her mother and her interest in science from her Army-officer father, was a successful writer of children's tales and romances, as well as of a number of attractive books on popular astronomy for young people. Mary Proctor, Richard Proctor's daughter, followed her father's lead as a popular writer and lecturer, chiefly in America, where she was known as 'the children's astronomer'.[43]

Among the younger women, Mary Orr, who married the solar physicist John Evershed, had her interest in astronomy nurtured in the British Astronomical Association and went on to become an excellent solar physicist – albeit as an 'obligatory amateur' – and an authority on the astronomy of Dante.[44] Agnes Clerke was, in her quiet way, sympathetic to the efforts of women amateurs. 'Her influence was inspiring to beginners of the science she so much loved.'[45]

Agnes Clerke took an optimistic view of the progress of women in the intellectual world. 'In the past female contributions to knowledge were valuable, it is true, but exceptional and unsystematic', she once wrote. 'Now, at last, they bid fair to become so serious and habitual as to be admitted without surprise, and appropriated without compliment.'[46]

It was to be some years, however, before women became 'habitual' members of the scientific community. Agnes Clerke's own position in the astronomical world, and that which made her uniquely successful, lay in the wise choice she had made of how to put her talents to best use. Allan Chapman sees her as 'really a "stateless" person insofar as she was neither a Grand Amateur, an assistant, a spouse, nor a lady of ample means, but a woman who stood outside the professional scientific community yet who made a living by studying its history and writing about it.'[47]

14 Revised *System of the Stars*

A new edition of *The System of the Stars*

As soon as *Problems in Astrophysics* had been reviewed in 1903 Agnes Clerke decided to bring out a second edition of her earlier book, *The System of the Stars*. The first edition of 1890 had been published by Longmans, but as that firm was not interested in a second edition Agnes Clerke arranged with Blacks, the publishers of her other two major books, to take it on.

'I want to make the book essentially of the twentieth century, retaining the old form yet little of the substance', Agnes Clerke told Gill, adding 'I shall be covetous of photographs, advice, information – nothing that you will be kind enough to tell me will come amiss.'[1] A few days later she wrote to Campbell at Lick, telling him also of her plan for the new edition, having first thanked him for his kind review of her *Problems* which had appeared in the last number of *Astrophysical Journal*. 'I do not venture to ask for loan of any valuable photos which you are reserving for separate and special publication, but may beg leave to use your diagrams from Bulletins.'[2] Campbell replied positively, saying that he hoped to be in a position to allow her to use the Lick nebula photographs.[3]

Agnes Clerke wrote in a similar vein to Pickering, asking for permission to reproduce certain sets of Harvard spectra among her full-page photographic illustrations.[4] Permission was granted, and the photographs sent immediately, accompanied by an invitation to the Clerke sisters to visit Harvard. But they were no longer young, and Agnes at least did not feel equal to it.

> My sister and I should think ourselves greatly privileged in seeing you and Mrs Pickering in your beautiful and famous home, and to my work

the advantage would be inestimable of consulting the archives of your great Observatory. But I fear the journey is an impossibility, and we must content ourselves with the help and sympathy accorded to us very generously from a distance.[5]

This was a pity; it would have widened Agnes' understanding of the travails of observational astronomy to have witnessed the Harvard women at work.

George Ellery Hale

Agnes Clerke also applied for material for the new edition of *System of the Stars* to George Ellery Hale. Hale, still only 30 years of age, was now Professor of Astrophysics at the University of Chicago and Director of the Yerkes Observatory, which he had been instrumental in founding in 1897. Agnes Clerke was already corresponding with him on the question of the identification of the numerous lines of unknown origin in the solar spectrum. Hale had obtained some marvellously detailed solar spectra at Kenwood, his original observatory, and at Yerkes. Agnes Clerke asked about his forthcoming list of chromospheric wavelengths, suggesting the possibility of identifying among them the spectra of rare gases, recorded in the laboratory by James Dewar (her friend from the Royal Institution) and his Cambridge associate George D. Liveing[6] but not yet published. She later informed him of further lists of accurate wavelengths published by the spectroscopist Edward C.C. Baly of University College, London.[7] Hale promised to send his list of solar wavelengths when they were ready, and explained that he had been fully occupied with setting up his new spectroheliograph at Yerkes.[8] This was the Rumford spectroheliograph which had taken much labour to complete but which more than fulfilled his expectations; his first satisfactory photographs were obtained in May 1903.[9] With this instrument Hale photographed the sun in the light of the calcium K line, and discovered the bright calcium clouds which he called 'flocculi'. These showed quite a different pattern on the sun from the hydrogen alpha picture. 'The photographs taken with the K line at different levels, and the dark structures photographed with the hydrogen lines, will possibly prove of some value', he wrote modestly to Agnes Clerke.[10] His letter is dated 27 May; she was therefore among the first to be told of his new brilliant success. He congratulated her on her honorary Membership of the Royal Astronomical Society, recently announced. He himself was soon to be honoured by that Society with its gold medal for his work in solar photography.

Examples of Hale's new spectroheliographs were sent to the Royal Astronomical Society, where they 'excited much admiration'.[11] Following his published account, Agnes Clerke wrote: 'I have been reading with profound admiration your statement of the results obtained with the Rumford spectroheliograph, illustrated by the most remarkable set of plates that have ever accompanied a solar investigation.' She went on to discuss the location of the reversing layer, as being, she thought, 'between the level of the production of the "floccules" and that of faculae. Should this be considered as proved, it would give a most valuable holding-ground for building up the successive layers of the sun's appendages.'[12] Here was Agnes Clerke once more with her tidy model of the sun's atmosphere. Hale was more cautious: 'I hardly think it [the reversing layer] could be strictly said to lie between the level of the protection of flocculi and that of faculae. . . .'[13]

Agnes Clerke's letter had followed Hale to California, where he went in December (1903) to make plans for his new proposed solar observatory on Mount Wilson. His detailed three-page reply shows that Hale took Agnes Clerke seriously as a sort of scientific consultant.

> Your kind letter of December 13 reached me here in California and was very welcome indeed. I am glad to know that the recent spectroheliograph results have proved of interest. If as I hope the method is still in an early stage of development, I think we may anticipate other interesting and perhaps novel results in the future . . . It is in the interest of such work as this that I am at present in California. I hope to establish a horizontal telescope of about 150 feet focal length on one of the neighbouring mountains, in case the conditions prove to be as favorable as our investigations seem to indicate. It is evident that a large image of the Sun will be required in the extension of the spectroheliograph work, and with the fine seeing I had here last June there should be no difficulty in using such an image.[14]

A month later he continued the story:

> I am establishing a small horizontal telescope on Mount Wilson, and
> hope to have it in operation quite soon. We are fitting up a very
> dilapidated old log cabin as a dwelling house and office. You may be
> interested to know that the library on Mt Wilson now consists of Miss
> Clerke's *History of Astronomy* and her *Problems in Astrophysics*,
> together with Spencer's *Synthetic Philosophy*, and a small volume of
> d'Annunzio's poems.[15]

She in turn hoped that his library would soon be reinforced by the
second edition of *System of the Stars*, which she proposed to change
'very considerably',[16] and reminded him that his review of the first
edition in 1891 'was among the few which had both a stimulating and a
corrective value'. She had read it again and had eliminated all that he
considered 'questionable or unsound'.[17]

For the new edition, Agnes Clerke asked Hale's permission to use
a set of the Yerkes spectra of various types of stars;[18] this he was
delighted to give. 'In fact, please feel at perfect liberty to reproduce any
of our photographs at any time without asking special permission. I feel
it a great honor to have them appear in your works.'[19]

Hale was far too busy at Mount Wilson to travel to London to
receive the Royal Astronomical Society's gold medal in person at the
February 1904 meeting. Agnes Clerke was there, and reported to him:

> The astronomers of England have thereby done their utmost to show
> their appreciation of your work, which has an importance transcending
> that of any isolated discovery in so far as it throws open an endless
> vista of future acquisitions to knowledge, many of them destined, so
> far as one can foresee and hope, to be realised by yourself.[20]

She sent 'best wishes for the success of your grandiose project for
a mountain observatory' from her sister and herself.

The second edition of *The System of the Stars*

The new edition, which came out in late 1905, contained twenty-seven
chapters, twenty-five of them updated from the first edition of 15 years

earlier, the extra two dedicated to new topics – spectroscopic binaries and stellar temperatures. Coming so soon after *Problems in Astrophysics*, the book made use of much of the same material.

So engrossed had Agnes Clerke become with the minutiae of spectroscopy, especially of binary and variable stars, that progress in the more classical topics received less than their due attention. This applied noticeably to the subject of stellar statistics from star-count surveys, especially that of Kapteyn based on the Cape Durchmusterung, which ought to have been of special interest to Agnes Clerke on account of Gill's connection with it. It is interesting to compare with hers the treatment given to this important topic by an astronomer of the old school, Simon Newcomb, whose book *The Stars, A Study of the Universe* was published in 1902.[21] Newcomb's unpretentious book gives actual star numbers and discusses the density of stars in space in the vicinity of the sun and the question of the increase in star numbers with magnitude, data that have a bearing on the subject of the structure of the stellar system. He concluded that stars were thinning out with distance, another step on the road to figuring out the shape of the Milky Way system. Agnes Clerke had read Newcomb's book when it came out,[22] but did not include the material in her own beyond a general updating of her chapter on 'the construction of the heavens'.

Another omission is any reference to the age of the universe, or to the conflict between astronomical estimates of the age of the sun with the much longer geological age of the earth. This, too, is discussed in Newcomb's book. *Problems in Astrophysics* also shunned the inconvenient but important question of the astronomical time-scale; Helmholtz' theory (1853) of the sun's contraction as its energy source, accepted by most scientists, was reported but not discussed in any depth.[23] All that Agnes Clerke has to say on the subject in her *History*, having given the accepted estimate for the duration of the sun's energy (18 million years or double this at most), was 'But this is far from meeting the demands of geologists and biologists'.[24]

The picture of the universe, presented in the first edition of the *System of the Stars* in 1890, still stood in the second edition of 1905. Vast numbers of nebulae, mainly spirals, had been discovered by Keeler at Lick Observatory in the interval, but this did not affect the general conclusion that they all belonged to one system. Indeed, a quite different idea was floated, that the spiral nebulae were solar systems in the making,[25] the result of tidal action between two passing stars according to the mechanism proposed by the geologist T.C. Chamberlin for the formation of our own solar system. But though Agnes Clerke conscientiously reported in the second edition of her book that 'the conviction is now dominant that they [the spiral nebulae] originated through some kind of repulsive action' between two sun-like bodies, she herself was obviously unconvinced, and omits completely the hypothesis of solar system formation.[26]

One important new observation was a spectrum of the Andromeda Nebula obtained by the German astronomer Julius Scheiner, showing absorption lines on a continuous spectrum. Scheiner saw it (correctly) as indicating that the nebula was 'virtually a cluster of sun-like stars', but Agnes Clerke – perhaps because it suited her convictions – preferred an earlier spectrum by the Hugginses, interpreted by them as showing faint emission lines, which would indicate a gaseous element.[27] She concluded that the question would have to be left in abeyance until better spectra were available.[28] Yet another explanation, Lockyer's familiar one, that the nebulae were made of meteoric dust, was reported without comment.

As far as Agnes Clerke (speaking for the majority of astronomers) was concerned, the 'system of the stars' (in the singular) of 1890 still ruled in 1905.

Illustrations

Agnes Clerke planned to complete the book in the course of the year 1904, and in January wrote to Campbell again, congratulating him on the Lalande Prize and reminding him of his conditional offer of copies of the 'most magnificent nebula pictures' from which the book would gain in value and interest.[29] A further request for a photograph of Nova Persei 1901 followed.[30] Campbell (who had been laid up for a long time following an accident in January 1903 and was still not fully recovered) had delayed replying until he should be able to give her a positive answer, but then wrote most generously:

> It is probable that I can accede, gladly, to any request you may make for copies of Professor Keeler's Crossley photographs; or, in fact, for any of our photographs, as I hope to have Professor Keeler's work published before next spring. A part of the money has now been raised, and I am going to San Francisco next week to make a final desperate effort to secure the remainder. As you probably know, I have to depend upon private generosity for funds to provide the illustrations for the Keeler volume.[31]

A week before this letter arrived, a worried Agnes Clerke accepted an offer from David Gill, who happened to be in London at the time, of Victoria telescope photographs which were already on their way from the Cape. Though 'the Crossley pictures would have been the most effective ornament to the chapter on nebulae', she felt it would be ungracious to decline those from the Cape. 'Yet I fear they are not likely to come up to the standard of your incomparable collection.'[32] A third photograph from the Cape specially requested, of 30 Doradus in the Large Magellanic Cloud,[33] declared by Agnes Clerke to be splendid and 'a novelty for the book'[34] arrived too late to be used. On the other hand, there was the compensation of a supply of striking celestial photographs by Max Wolf of Heidelberg, who had been supplying her with photographs for many years. His survey of nebulae (galaxies), begun in 1901, revealed multitudes of these objects in groups or clusters. Agnes Clerke was in close correspondence with Wolf at this time, and was fortunate in being able to use his most recent photographs of the Andromeda Nebula, M33 and the Cocoon Nebula (obtained in 1904) (Figure 14.1). One illustration was retained from the first edition – the pull out frontispiece map of the Pleiades by the brothers Henry.

There were also several reproductions of relevant spectra, though these, as in her other books, were not always helpful. Agnes Clerke was undoubtedly more at home with words than with practical matters. She published the spectra exactly as they were supplied to her, some with identifications, others with none. Her contemporaries were perhaps familiar with them already, while non-specialist readers had to be satisfied with her verbal explanations.

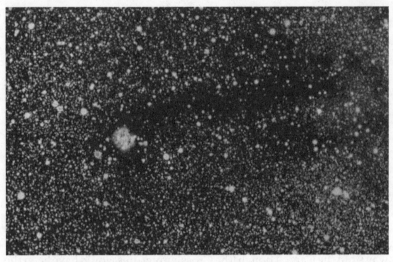

Figure 14.1 The Cocoon Nebula in Cygnus, by Max Wolf.

Copies of the book went immediately to Gill, to Hale and to Campbell, who was looking forward to reading it.[35]

Nature attacks again

Agnes Clerke cannot have looked forward with any pleasure to the review in *Nature*, which appeared after a delay of several months. The reviewer was again Gregory, who regrettably repeated the same prejudiced remarks about the female author which had greeted *Problems in Astrophysics* two years before. His report,[36] four pages long, opened with a French proverb, translated as meaning that

> the intuitive instinct of a woman is a safer guide to follow than her
> reasoning faculties . . . and though in these days it is considered
> ungracious to make these suggestions, evidence of its truth is not
> difficult to discover in most literary products of the feminine mind. It
> is no disparagement to Miss Clerke that even she shares this
> characteristic of her sex.

Gregory repeated his earlier strictures, blaming the author heavily for favouring Sir William and Lady Huggins at the expense of Lockyer, not only in the matter of Lockyer's meteoritic theory (an old debate), but also in the discussion of stellar spectra and the temperatures of stars and nebulae. It cannot be denied that, while giving a fair showing to Lockyer's explanation of the spectra of hot stars in terms of dissociation, she came down, as always, in favour of the Hugginses. 'Sir William and Lady Huggins demonstrated in 1897 that the calcium lines H and K owe their preponderance in the sun not to extreme heat but mainly to the inconceivable rarity of the emitting vapour',[37] she wrote in the new edition, and 'the hypothesis of its [calcium] dissociation in the sun thus remains unverified'.[38] This and other examples were cited by Gregory to show how Agnes Clerke presents only one side of a case as if it was 'the last word'. 'Science', he wrote, 'is not a persuasion in which personal opinion has to be respected ... We may admire Miss Clerke's literary skill ... but, at the same time, we may be permitted to recollect that she is a bibliographer rather than an observer, and therefore her works, be they ever so distinguished as literature, need only be regarded as narratives by a spectator, when the weights of the results and conclusions recorded in them are being decided.' Nevertheless, he paid tribute to her virtues: 'we can forgive the occasional florid style when we remember the vast amount of reading and careful analysis involved in the preparation of a volume of this kind. The work is so good that every student of astronomical physics must be familiar with it, and every astronomical library must include it.'

Gregory's main complaint – that Agnes Clerke was a partisan of the Hugginses – is unlikely to have troubled her devotees, who were aware of and admired her friendship with that revered couple. It was a pity that he saw fit to preface his review with remarks which have been seen as an example of anti-feminist attitudes in Victorian Britain.[39]

Gregory took a more positive – indeed an approving – view of Agnes' next book, *Modern Cosmogonies* (1905), published in the same year, which he recommended as containing much to interest the general reader.[40]

Physics and cosmogonies

The book *Modern Cosmogonies* originated in a series of articles on theories of the origin of the earth contributed in 1904–5 to the popular journal *Knowledge*, now renamed *Knowledge and Illustrated Science News* under its new editor, Baden-Powell. As soon as the series was finished it was published in book form by A. and C. Black.[41] It was a popular account of the structure of the astronomical world and the history of the various models of the universe from the Greeks to the time of writing.

Short though it was (fewer than 300 pages) by comparison with her three major works, it was a remarkable achievement in that it was not entirely about astronomy but touched on many fields of physical and biological science. The formation of the solar system, the nature of stars, the universal forces of nature, the ether, the geology of the earth, the evolution of life, the problem of keeping the stellar system from collapsing – all were discussed. Her readers also encountered the recent discoveries of radioactivity, and such unfamiliar ideas in physics as the undulatory theory of gravity and the interchangeability of mass and energy.

Modern Cosmogonies emerged in parallel with some articles by Agnes in the *Edinburgh Review* about new and much-talked-of scientific and technological discoveries at the turn of the century. Agnes Clerke would have kept abreast with these developments chiefly through the Royal Institution lectures. One of these subjects, described in an article entitled 'Ethereal telegraphy' in October 1898, was wireless (radio) transmission.[42] In 1896, Guglielmo Marconi, who had begun his experiments in Italy a few years earlier, arrived in England to continue his work. He carried out successful tests in 1897 on Salisbury Plain, sending signals from balloon-borne aerials over distances of several miles – demonstrations that attracted considerable publicity and interest and were the subject of a lecture at the Royal Institution the same year.

The concept of radio waves, or Herzian waves, as they were then called, would not have been new to Agnes Clerke, who, back in 1889, only one year after Hertz's discovery, would have heard Oliver Lodge's lecture at the Royal Institution and witnessed his demonstration of discharges from a Leyden jar. After Hertz's untimely death in 1894 Lodge, himself a pioneer in the field, again lectured with demonstrations on Hertz's work at the Royal Institution. In her essay, Agnes Clerke reviewed a list of publications including the works of Hertz in English translations (1893) and Lodge's *The Work of Hertz and Some of his Successes* (1894). As was her wont, she treated the subject historically, beginning with Faraday, through James Clerk Maxwell to Hertz and Marconi.

The new century was also a time of sensational revelations in physics. 'It cannot be denied that the phenomena of radioactivity are subversive of ordinary ideas in physics', wrote Agnes Clerke in an essay on 'The revelations of radium' in the *Edinburgh Review* for October 1903.[43] In the spring of that year Ernest Rutherford, the scientific 'lion of the season', had delivered an address to the British Association for the Advancement of Science at Southport in anticipation of which the newspapers had 'suddenly become radioactive'.[44] Rutherford and Frederick Soddy, working in Montreal, had performed their brilliant experiments, reported in a succession of papers published in 1902, which revealed that radioactive elements must undergo spontaneous transformation within the atom – an entirely revolutionary idea at the time. Rutherford was elected a Fellow of the Royal Society – of which Sir William Huggins was now President – during that visit to Britain.

At around the same time Pierre and Marie Curie visited London when Pierre, speaking in French, addressed the Royal Institution (June 1903) on the subject of 'Radium'. While Pierre lectured Madame sat next to Lord Kelvin among a throng of the country's most eminent scientists in the crowded hall.[45] It is quite possible that Agnes Clerke, one of the Institution's privileged circle, met the Curies on this occasion. In his lecture Curie described the heat-generating property of radium preparations. This new 'mystery of radium' was demonstrated, as Agnes Clerke tells in her essay, 'with the aid of the fine refrigerating equipment of Professor Dewar'. The Curies were feted by the country's senior scientists, who included some whom Agnes Clerke knew well at the Royal Institution – Dewar, Ayrton and Sylvanus Thompson, and of course the Hugginses – and were afterwards entertained to a glittering dinner at which Marie Curie sat next to Lord Kelvin.[46]

Agnes Clerke was fascinated by what she heard, and got down straightaway to study Madame Curie's 'bewildering discovery' and to write her article for the autumn number of the *Edinburgh Review*. The publications listed for review in her essay included Rutherford and Soddy's 1902 papers in *Philosophical Magazine*, the 1903 Wilde Lecture on *The Atomic Theory* delivered by the mineralogist Frank Wigglesworth Clarke to the Manchester Literary and Philosophical Society, a German monograph on radioactivity and the recent papers by the Hugginses on their observations of the spectrum of radium.[47]

The article opened with John Dalton's axiom 'No man can split an atom' and went through the experimental evidence whereby this idea came to be slowly transformed: William Crookes' experiments with vacuum tubes from 1873 onwards, the discovery of cathode and Röntgen rays, J.J. Thomson's discovery of the electron (demonstrated at a Royal Institution lecture in 1897), the discovery by Marie Curie ('this gifted Polish lady') of polonium and radium (citing her Paris thesis, now translated into English) and Pierre's observation, the subject of his London lecture, of the heat-generating power of radioactive material. The conclusion, that the atom was no longer the ultimate elementary particle, was not easily accepted by some.[48] Agnes Clerke wrote to Gill in September:[49] 'I have written an article on Radium for the October number of *Edinburgh Review*, and have consequently got a particularly vivid impression upon the brain of the "perpetual flux of things" (as some old philosopher used to phrase it). Madame Curie's bewildering discovery has greatly increased the electrical tension of the scientific atmosphere, and one never knows in mentioning the subject where one may tread on a "live wire".' Agnes Clerke ended her article, as she so often liked to do, with a classical quotation: 'Of atoms, as of men, it may be said with truth, "Quisque suos patitur manes".'

For readers of the *Edinburgh Review* Agnes Clerke's article, appearing in October 1903, could not have been more timely. In November the Curies were awarded the Humphrey Davy Medal of the Royal Society. Marie being unwell, Pierre travelled to London on his own and was the guest of the Hugginses (Sir William was President of the Royal Society) in their observatory-home at Tulse Hill.[50] And at the end of the year the Curies and Henri Becquerel were awarded the Nobel Prize.

Agnes Clerke was to return to the subject of radioactivity in 1906, when Rutherford published his famous book *Radio-active*

Transformations, with a report on it in the *Edinburgh Review*. This time she chose not to repeat the detailed historical account of the subject that she had given in her earlier article. Instead, under the title 'Old and new alchemy',[51] she coupled her discussion of the transformation of the chemical elements through radioactivity with reflections on the ancient dream of the alchemists to turn lead into gold. The additional books reviewed were five works on alchemy and hermetic philosophy published at various dates from 1742 to 1897, two each in German and French and one in English. (One may well imagine that her omniscient astrological friend Richard Garnett helped to supply this esoteric list.) The result was a delightful essay, in which we meet Egyptians, Babylonians, Pliny the Elder, Saint Augustine, and one James Price, a Scottish alchemist. This article, her fifty-third and last in the *Edinburgh Review*, appeared in January 1907, the very month of her death. That same month, too, Ernest Rutherford returned to England to take up the Chair of Physics at Manchester.

At the close of 1905, after 3 or 4 years of extraordinarily intense work carried out in spite of failing health, Agnes Clerke had achieved her aim of presenting an up-to-date panorama of astronomy to twentieth-century readers: the fourth edition of her *History* (1902), her new *Problems in Astrophysics* (1903) and the second edition of *The System of the Stars* (1905), with *Modern Cosmogonies* (1905) thrown in as a bonus.

Death of Ellen

But the year 1906 was to bring to Agnes Clerke the greatest sorrow of her life – the unforeseen death of her sister and lifelong companion, Ellen. She died at the age of 65 on 2 March 1906, after only two days' illness, following a cold that developed into pneumonia.

Ellen Clerke's activities as a journalist have already been mentioned. Her regular *Dublin Review* features on foreign geography and on social and religious affairs abroad continued until 1901; her later contributions were confined to shorter book reviews. Her last major article in that journal, 'Catholic Progress in the Reign of Victoria'[52] was written after the death of the Queen in 1901. Like her sister's article on astronomy at the time of the Queen's Jubilee in 1897, Ellen's was an unabashed litany of praise of the Empire. The Catholic Church, she pointed out, was being more kindly tolerated at home and was spreading throughout the British dominions. In this sentiment Ellen was at one with her patron Cardinal Herbert Vaughan, who in a Pastoral letter delivered at the opening of the new century in January 1900 spoke of the 'divine mission' of the Empire in spreading the Christian message across the world.[53] She narrated – with the help of copious statistics on population, clergy, schools and religious houses – how the British Catholic world under Victoria had emerged from 'three centuries in the catacombs', so that being a Catholic no longer carried a 'social stigma'. She did not avoid the distressing subject of the Irish Famine, which had cast a cloud over Victoria's reign, but was able to give it a positive interpretation by recounting the impact of Irish emigration on the colonies, and the success of the Irish as pioneers in the extension of both the Church and the Empire 'in the Great West and the Great South'.

Ellen's work for the *Tablet*, which continued until the end of her life, was mainly concerned with European affairs. Her obituarist in that journal praised her 'quick and eager sympathies' with Italian problems and her 'intellectual mastery' of German politics.[54]

However, Ellen's preferred taste was literary. She had a great love for Italy, and contributed several articles on Italian literature and culture to various magazines besides the *Dublin Review* – the *Edinburgh Review, Fraser's Magazine, The National Review* and *Temple Bar.* She published two books of poetry, *The Flying Dutchman and Other Poems*,[55] a collection of her own original compositions, and *Fable and Song in Italy*,[56] a set of essays on Italian romantic poetry illustrated by her own translations rendered in the original metre. The latter volume, dedicated to Richard Garnett 'in grateful recognition of encouragement and advice', was well received, wrote Agnes to Lady Gill on sending her a copy. One critic praised Ellen's success 'in the double capacity of poetical translator and of literary historian'.[57] The poets in question were chiefly the fifteenth-century Ariosto and Boiardo, but also included the modern Allesandro Manzoni and traditional Italian folksongs.

A serious contribution to Italian studies was her part in Richard Garnett's *History of Italian Literature* (1898),[58] which quoted many of her translations. Late in life she wrote her one and only novel, *Flowers of Fire* (1902),[59] a romance set in Italy and Poland at the time of the Polish insurrection of 1863. The principal characters are two sisters and their patriotic Polish lovers who endure much hardship for their cause including prison and exile. The story ends happily with a double wedding. The action includes a visit to the observatory on Mount Vesuvius and a vivid description of a volcanic eruption (the Clerkes had witnessed the eruption of 1872 and probably also visited the observatory).

Ellen Clerke is also known to have contributed (in Italian) a long series of stories to a Florentine journal.[60] A full account of her life's varied work has yet to be written.

15 Cosmogonies, cosmology and Nature's spiritual clues

Nature's spiritual clues

Margaret Huggins, commenting on Agnes Clerke's *Modern Cosmogonies*, claimed that it was 'not only history, but a work of philosophical thinking and of imaginative insight of a very high order'. 'Where else', she asked, 'is shown in recent philosophical writing such vision and faculty divine for seizing and pointing out the reasonable spiritual clues, set in what we call Nature – clues helping to sustainment of soul in the midst of the majestic mysteries surrounding us.'[1] Years before, referring to *The System of the Stars* (1890), she had admired her friend's reverence for the Deity at a time when 'it has become much a fashion to be really afraid to even mention the word God when science is concerned'. In that book, Agnes Clerke had referred to 'the vision of a Higher Wisdom', brought about by the study of astronomy, and to 'the sublime idea of Omnipotence, to which the stars conform their courses while "they shine forth with joy to Him who made them"'.[2]

Such pious sentiments were common enough in the writings of an earlier generation. Mary Somerville in the Introduction to *The Connexion of the Physical Sciences* (1834) refers to 'the goodness of the great First Cause, in having endowed man with faculties by which he can not only appreciate the magnificence of his works, but trace with precision the operation of his laws'.[3] John Herschel mentions 'the Master-workman with whom the darkness is even as the light'.[4] Entirely practical handbooks, with no philosophical content, such as W.H. Smyth's *Cycle of Celestial Objects* (1844) were not without their

recognition of the Creator. Smyth refers to 'the whole firmament with
its countless glorious orbs which are . . . individual constituents of one
Majesty of Creation'. Popularisers of science, writing at an elementary
level for mass readership or for children, frequently presented the
marvels of the natural world as a demonstration of the wisdom and
power of God. Authors of this genre whose motives were moralistic
were numerous and their works much in demand.[5] Agnes Clerke,
however, does not fall into this category, which is why Bernard
Lightman in his study of nineteenth-century popularisers of science
comments: 'Ironically, Clerke, the new style populariser, perpetuated
in her astronomical work the older, natural theology tradition that was
so important in English thought previous to Darwin.'[6] In fact, Agnes
Clerke, by education and upbringing, was not part of the English tradition. Neither was she reverting to the earlier convention with these
religious allusions. She wrote from serious conviction.

St Thomas Aquinas

Agnes Clerke, as her various essays show, was well versed in the history
of science and philosophy, and in the intellectual thinking of scholars
throughout the ages. A devout and dutiful Catholic, she would have
conformed with the teachings of her Church on such issues. Pope Leo
XIII, successor in 1878 of Pius IX, began his reign with an Encyclical[7]
calling for a renewal of the study of theology and a return to the teachings of the great thirteenth-century philosopher and theologian, Saint
Thomas Aquinas, whom he declared a Doctor of the Church. To this
end schools of theology staffed by eminent scholars were set up in the
Catholic University of Louvain in Belgium, in Rome and elsewhere.[8] It
was the beginning of a renaissance of Thomist studies, which remain
today an active branch of academic theology in many centres of
Christian learning.

Regard for Thomas Aquinas was not confined to scholars. It
extended to the entire Catholic faithful. Thomas Aquinas, as well as
being a great thinker, was a great saint, a man of immense prayerful-
ness, humility and simplicity, canonised in 1323. The liturgy which he

composed for the feast of Corpus Christ in 1264 includes some moving hymns, such as the motet *Panis Angelicus*, which are familiar well outside the Roman Missal and Breviary.[9] Prayers composed by St Thomas Aquinas are found in most Catholic devotional books.

Thomas, a Dominican monk, was a student of another great Dominican scholar Albertus Magnus at the University of Paris. It was the time when western countries were re-discovering Greek scholarship, maintained in the Islamic countries when it was all but lost in Europe. The reconciliation of Aristotle's world picture with Christian theology, a major step forward in European science, was achieved by the scholastics at Paris, chiefly Thomas Aquinas. The modern historians of science Michael Hoskin and Owen Gingerich[10] point out that at the University of Paris, the great intellectual centre of Christendom in the thirteenth century, there had been a traditional division of learning between the liberal arts (which included mathematics and astronomy) and the higher faculties (medicine, law and especially theology). The synthesis between the teachings of Aristotle concerning the natural world and Christian theology achieved by Thomas Aquinas meant that astronomy now found its place among the higher branches of learning.

Among Thomas's important works was a commentary on Aristotle's book on the heavens, *De Coelo et Mundo*. A 'magnificent edition' of this commentary (thus described by J.L.E. Dreyer, the eminent astronomer and historian of astronomy, in his *History of the Planetary Systems from Thales to Kepler*[11]) was published in Rome in 1886 and would most certainly have been read by Agnes Clerke. Her own *Modern Cosmogonies* came out just ahead of Dreyer's book. Dreyer, in his chapter on Medieval Cosmology in that *History*, gives the core of Thomas Aquinas' *De Coelo et Mundo*.

> It is a commentary on Aristotle's book on the heavens, and the spirit in which it is written shows the vast strides from darkness towards light which had been recently made. Though Aquinas was deeply convinced that revelation is a more important source of knowledge than human reason, he considered both to be distinct and separate ways of finding truth; and in expounding Aristotle he therefore never lets himself be disturbed by the difference between his doctrine and that of the Bible, but assumes both to be ultimately derived from the same source.[12]

Thomas Aquinas' commentary on Aristotle was only a small part of his immense volume of writings. His best-known and most important works were his profound treatises on theology, *Summa Theologica* (Summary of Theology) and *Summa Contra Gentiles* (Summary against the Gentiles), an exposition of the Christian faith addressed to non-believers and sceptics. One of the aims of the latter was 'to show that the Christian faith rests on a rational foundation and that the principles of philosophy do not necessarily lead to a view of the world which excludes Christianity either implicitly or explicitly'.[13]

Starting from the experience of our senses, Thomas Aquinas deduced a number of 'ways' or proofs for the existence of God.[14] Motion or change demands the existence of a first or supreme unmoved mover. Cause and effect lead back in a chain to a first cause. Material things, though unconscious, behave with an 'end' or purpose which points to the existence of an extrinsic intelligent author.[15] Reason is not enough, however, without revelation – the revelation that God made the world and made it rational. One writer puts it:

> The Christian doctrine of creation . . . tells us that the world was created by a rational God and that we can expect therefore to find in the world regularities and patterns reminding us of God. The same doctrine makes it clear that we cannot expect to predict those patterns *a priori* from pure reasoning since God is free and could have created a world with very different patterns.[16]

On the relation between science and theology, a modern Catholic philosopher elucidates:

> The scientist has to act on the premises that an external world exists, that it is orderly, and that the mind has the capacity to grasp the order that is there. Further acts of faith are demanded for anyone to become committed to the scientific enterprise, to learn the current state of the discipline, and to advance towards new discoveries . . . Conversely, religious faith depends to some extent upon reason. Revelation could not be made but to a rational being, for a brute animal could not grasp the meaning or credibility of God's word.[17]

Agnes Clerke expressed many of these ideas in her introduction to the Hodgkins Essay in 1901, a work intended by the terms of its foundation to deal with the place of the Creator in science (Chapter 11). She wrote:

> Law formulates intelligent purpose; and the laws of nature are an expression of the Will of God. In tracing them out, man seeks, more or less consciously, the infinite: and his capability of tracing them is derived solely from the analogy of his mind with the Creative Mind, which designed a universe assumed by the necessities of thought to be a 'cosmos' – an orderly arrangement, such as his faculties can apprehend. Were it unplanned, or planned according to a method incomprehensible by human reason, science would have no *locus standi*: life should be conducted on purely empirical principles. As a fact, however, we find the world essentially intelligible; and by striving to enlarge the limits of its intelligibility, we promote the purpose of the Creator in placing us there, and, following in the track of His primal conceptions, bring our inchoate ideas more and more into harmony with them.[18]

Though she does not use formal philosophical terminology in her cosmology, Agnes Clerke appears always mindful of St Thomas' teachings. Her very use of the words 'reason and revelation' shows this. She would have obtained guidance from Catholic theologians or in the Catholic journals – the *Dublin Review* and the *Tablet*. Turning back the pages of the former she would have found a scholarly essay entitled 'Arguments for the Existence of God' published in 1885,[19] and another in the same number which clarified the separate roles of science and the Church, from the Catholic viewpoint.[20] A third, published in the *Dublin Review* next to her own article on the Cape (1889), was a robust answer to the philosophy of the agnostics, Huxley, Tyndall, Spencer 'and other illuminati'.[21]

Spiritual clues in cosmology

Agnes Clerke's 'spiritual clues' are mostly to be found in *The System of the Stars*. The sidereal universe as revealed by observation 'cannot be regarded', she claimed, 'as in any true sense infinite'. It 'bears glorious witness to the power and wisdom of the Almighty Designer: yet it has limits, and for that reason it is a fit subject for the exercise of limited understandings.'[22] It is finite, because 'human reason would otherwise be totally incompetent to deal with the subject of its organisation'.[23] This statement presupposes that human reason is competent – as St Thomas has it – to learn about the universe from observation, a situation that Agnes Clerke might have considered inconsistent with a universe extending to unreachable infinity in space and time. The universe was created and also changes: 'Are the stars subject to growth and decay? We might almost as well ask, Are they subject to the laws of Nature? . . . We are perfectly assured, both from reason and revelation, that a time was when they were not, and that at some future date they will not be.'[24] She dismisses the notion of 'a contemporaneous universal origin' for all the stars and rejects the idea of the universe as 'a mechanical workshop, turning out objects by the score in blind pursuance of an original intelligent arrangement'.[25] This last remark hinges on the place of the Creator in the world. Thomas taught that: 'The preservation of things by God does not take place by some new action, but by a prolongation of that action by which he gives existence; and this action is without change or time.'[26] The idea of God devising a set of laws, switching on the universe and letting it take care of itself, is not valid by Thomas' reasoning.

On the scientific level, Agnes Clerke subscribed broadly to the nebular hypothesis, which saw nebulae or nebulous material of some sort as the antecedent of stars and of planetary systems. She was herself familiar with the ideas of Kant and Laplace,[27] which were also discussed in the theological article in *Dublin Review*. But where did that primeval material come from? 'It is not enough to refer stars to nebulae, while nebulae themselves are unaccounted for', she wrote. 'Prenebular' theories cannot give the ultimate answer. 'Such efforts to get nearer to an absolute beginning illustrate the incapacity of the human mind to rest finally in any material conception.' They merely pile 'Ossa upon Pelion in the shape of hypotheses, vainly hoping that, with the last and latest, the empyrean may at last be scaled . . . science can only become truly rational when the count of all things is reached in an Intelligence akin to, yet infinitely transcending, its own'.[28] Agnes Clerke here uses Thomas' argument of the First Cause, the Christian doctrine that Man is created in God's likeness, and also Thomas' teaching that it is right for mankind to use his reason to investigate the world.

To investigate the world, however, demands a belief in the universality of the laws of science, 'a law of order, the same always and everywhere'.[29] This, of course, is the starting point for every cosmologist.[30] In addition, these laws have been implanted by God. 'There is no such thing as chaos [by which one assumes is meant absence of law] in the sidereal world or outside it. For chaos is the negation of law, and law is the expression of the Will of God.'[31] This is in line with Thomas's teaching, which rejects the possibility of working out the laws of nature a priori from pure reasoning. Perhaps Agnes Clerke was herself guilty of a certain amount of a priori reasoning in her zealous insistence on the one-system universe: she regarded as unthinkable a universe of infinite dimensions in space and time.[32]

The astronomer's field of work is the observable universe, no more. 'With the infinite possibilities beyond, science has no concern.'[33] Here, I believe, Agnes Clerke is not referring to other possible undetectable universes in infinite space beyond the Milky Way (as Simon Newcomb allows in a popular book published about the same time: 'That collection of stars which we call the universe is limited in extent ... This does not preclude the possibility that far outside of our universe there may be other collections of stars of which we know nothing'),[34] but to aspects of the world which are beyond the reach of science.

Agnes Clerke believed (to quote one of the *Dublin* reviewers) in 'the exceedingly limited range of the human intellect'. 'We shall surely not wander far from the truth', she wrote, 'by recognising our inability to penetrate all the depths and complexities of Infinite Design'.[35] This she expresses in poetic style at the end of *The System of the Stars* (following her summary of the future fate of 'the millions of bodies that once were suns ... inert, lifeless, invisible').

> That is all we know; at the brink of the ocean we pause, helpless to sound its depths, or number the modes of its manifestations, or predict the tasks of renovation or preservation committed to it. We can only recognise with supreme conviction that He who made the heavens can restore them, and that when the former things have passed away, and the scroll of the skies is taken out of sight 'like a book folded away' a

'new heaven and a new earth' shall meet the purified gaze of recreated man.[36]

The moving image of the skies as a scroll being folded away comes from St John's Apocalypse.[37] It is the same one used by J.P. Nichol in what may have been the first book on astronomy that Agnes read as a child. Nichol, an ordained minister of the Church of Scotland before becoming an academic astronomer, was a devout believer in the Almighty. He, too, reflects on a time when the sun and stars cease to shine.

> Nay, what though all should pass? What though the close of this epoch in the history of the solar orb should be accompanied, as some with strange fondness have imagined, by the dissolution and disappearing of all these shining spheres? Then would our Universe not have failed in its functions, but only been gathered up and rolled away, these functions being complete.[38]

It is a touching thought that Nichol's charming book, preserved by Agnes throughout her life, had made such an indelible impression. Her copy of it is among the surviving relics of her childhood.

The age and evolution problem

A more difficult problem than cosmology for many Christian believers was the question of evolution. Geological evolution made the earth considerably older than the Bible, literally interpreted, would make it. Darwin's theory of biological evolution was an even greater obstacle, as it appeared to contradict the Biblical account of the origin of human-kind. In the former case, the Catholic Church had little trouble, making it clear that the Scripture was a narrative, not a scientific treatise. As to biological evolution, Agnes Clerke, surprisingly, devoted a chapter in *Modern Cosmogonies* to the origin of life, that is, to the question of how life in its primeval form arose from non-organic matter. She quoted Huxley's view that such a transformation occurred in a remote period in the past and also described the fossil evidence for very early types of primitive life. But, in the arrogant tone of which she was unfortunately capable, she branded as 'absurd' any notion that life at the start could have sprung from non-living material by some special configuration of molecules. 'Life is a gratuitous gift of the Creator', she had earlier stated in her review of Wallace's book. On the other hand, she had no intellectual difficulty with the doctrine of evolution through an ascending sequence of animal forms, regarding humankind as a being 'cast in a diviner mould'.

This one foray by Agnes Clerke into a complex subject in which she had no expertise was out of character. Richard Gregory's strictures would not have been misplaced in this instance (though in fact, as already mentioned, he found no fault with *Modern Cosmogonies* in his *Nature* review). One assumes that she was motivated by a desire to counteract as far as was in her power the spread among the general public of an agnostic view of science.[39] This focussed in particular on biological evolution and the work of the well-known agnostics (and violently anti-Catholic) Thomas Huxley, John Tyndall and Herbert Spencer[40] (though all three were now dead). Many of her readers no doubt, with Lady Huggins, applauded her stand.

16 Last days and retrospect

Arbiter and peacemaker

Agnes Clerke in her sixties had become a sort of mother figure among astronomers, tactful, kind, helpful. The sour comments of R.A. Gregory in *Nature*, about which she never complained, appear to have been the only shadow on her blameless life. She kept her balance in the dispute between Huggins and Lockyer without offending either, and subscribed to the superiority of the American spectroscopists without offending the Hugginses. A glimpse of her in the role of peacemaker is found in her correspondence in 1904 with W.W. Campbell of Lick, a man who in his youth was not afraid to stand up for himself against acknowledged 'authorities' (Huggins being one) if he knew his work was better.[1] There was a long-running difference of opinion between him and Huggins about the spectra of the Orion Nebula and the Orion trapezium stars. He also challenged the Hugginses' view, repeated in their recently published *Atlas*, that the Orion Nebula was variable. Campbell was now returning to the subject, and in a less aggressive mood than formerly sent a copy of his manuscript to Huggins before publishing. He placed the matter before Agnes Clerke, explaining that the reason for the differences between his observations and those of Huggins over the years was instrumental.[2] Agnes Clerke replied that 'no-one, least of all Sir William Huggins, will be likely to misconstrue the spirit of entire loyalty to science in which it [i.e. your investigation] has been conducted'.[3]

She was again invoked in a discussion concerning Hale's invention of the spectroheliograph and Henri Deslandres' claim to have thought of the same idea independently. The question of priority in discovery and invention was taken very seriously in the nineteenth century. One has only to think of the recriminations when astronomers in England failed to discover the planet Neptune after John Couch Adams had calculated its place and the honour went to Urbain Leverrier of Paris, whose prediction had been followed up without delay at the Berlin Observatory. Another example was the discovery of a means to observe solar prominences without a total eclipse by the use of a spectroscope, first carried out successfully in India by the French astronomer Jules Janssen on the day following the eclipse of 1868. Back in Britain, Norman Lockyer was able to show that he had hit upon the same idea earlier than his rival, though without the opportunity of trying it out. The result was a joint celebration in the form of a commemorative medal for each of the astronomers.

The argument about the invention of the spectroheliograph was similar. Hale was a friend of Agnes Clerke's since 1891, when he attended the Cardiff Meeting of the British Association and showed his spectroheliograms in Britain for the first time. After that meeting Hale was astonished to read a report of a similar invention to his own spectroheliograph made by Henri Deslandres, the Director of Meudon Observatory. He went to Paris immediately and saw Deslandres' instrument for himself. It was different from his. Hale himself has recounted the chain of events:

> About this time Deslandres introduced at the Paris Observatory his velocity spectrograph, which permits the motion in the line of sight of the calcium vapor at various levels to be measured on photographs of the H or K line in successive sections of the sun. Evershed in England soon constructed and systematically used a spectroheliograph, and in 1893 Deslandres also began work with a spectroheliograph, which he employed for photography with the calcium lines and with some of the narrower dark lines.[4]

In the third edition of her *History*, published in 1893, Agnes Clerke gave a faithful account of Hale's work, correctly referring to Deslandres as 'carrying on at the Paris Observatory the novel work initiated by Hale'[5] in connection with prominence photography in the ultra-violet. However, it later became widely accepted – and not denied by Deslandres himself – that the original Paris instrument was essentially the same as Hale's. Agnes Clerke did not have close contacts with Deslandres; she used one of his spectroheliograms in *Problems in Astrophysics*, which she acknowledged in the Preface to the book, but does not appear to have been on letter-writing terms with him.

Unfortunately, in the next and last edition of the *History* (1902), Agnes Clerke modified her account of Hale's work by saying that 'most of the novel researches on the sun's surface were, by a remarkable coincidence, pursued independently and contemporaneously, by M. Deslandres, of the Paris Observatory'.[6] The same interpretation was perpetuated in *Problems in Astrophysics* (1903) where she wrote: 'Their [solar prominences] imperfect visibility upon the disk greatly restricted their observation, until Professor Hale and M. Deslandres almost simultaneously invented a method for spectrographically recording them';[7] and in the second edition of *The System of the Stars* (1905) (to which Hale contributed his latest stellar spectra), where the two astronomers are recorded as having 'independently adapted' the double slit method of solar prominence photography.[8]

It was no doubt annoying for Hale to find his priority in the invention of the spectroheliograph questioned more than once in this way. He made his feelings known to Margaret Huggins who passed them on to Agnes Clerke and told Hale: 'Many thanks for your letter re the Deslandres pretensions. I think I quite understand, and you may trust me to be very discreet in talking over matters with Miss Agnes Clerke. Meanwhiles, trust me.'[9] She took the opportunity of speaking to Agnes Clerke:

> I want to tell you what of course you must have guessed, that I
> faithfully kept my promise to you, and made an opportunity to speak
> fully to Miss Clerke about giving you your due and proper statement –
> without wrong credit being given to M. Deslandres . . . I had her
> assurance that for the future you should have justice; and that she had
> certainly done you justice in certain Encyclopaedia articles. I felt quite
> satisfied with my little diplomatic mission.[10]

The outcome was a contrite apology to Hale from Agnes Clerke:

> I have taken careful note of your Reply to M. Deslandres . . . and
> perceive with distress that, in my books, I have inadvertently failed to

emphasise the originality and importance of your invention of the spectroheliograph. I need scarcely say that nothing was farther from my thoughts than to undervalue in the slightest degree your splendid work, the fame of which is world-wide; but I did not just then grasp the full bearings of your start with what has been called the 'monochromatic telescope'. However, I have made sure of setting things right should the sale of the volumes in question give me that opportunity; and I have meantime, in a sketch of astronomical history for the new edition of *Encyclopaedia Britannica*, given you the unqualified credit in the matter that belongs to you.[11]

Hale's response was characteristically generous. 'I need hardly tell you that your letter of August 18th has gratified me exceedingly. There is nothing I dislike so much as the discussion of questions of priority, and for this reason I never spoke to you about the statements in your books as to the first application of the spectroheliograph.' He went on to recall that when the Rumford Medal was bestowed on him in 1903, the Chairman of the Rumford Committee had advised him to write to Agnes Clerke about it. 'I disliked the question so much, however, that I have never felt inclined to raise it.' The letter illustrates Hale's modesty and high-mindedness, and also indicates Agnes Clerke's influence (as well as Lady Huggins'!) in the scientific community. Unfortunately, the intended amends in the *Encyclopaedia* article were not destined to be made. Agnes Clerke died before she had a chance to change the relevant lines in which the photography of prominences is described as having been realised by Hale and 'simultaneously' by Deslandres. Hale could easily have put the record straight in his own article on the spectroheliograph in the same edition of *Encyclopaedia Britannica*, but did not choose to do so, confining himself to describing his newest instrument at Mount Wilson.

The episode in no way affected the warmth of Hale's relations with Agnes Clerke; their correspondence on the subject of sunspot spectra continued, with Hale sending her the proofs of his forthcoming paper asking if she would notice it in one of the British journals. This review, published in the *Observatory*, was to be her very last piece of writing.

Encyclopaedia Britannica eleventh edition

The eleventh edition of *Encyclopaedia Britannica* was a major revision of that compendium of knowledge which had attained enormous popularity following the publication of a cheap reprint of the ninth edition in 1898. That edition, to which Agnes Clerke had contributed many years earlier, had been expanded by the addition of extra volumes to form the tenth edition in 1902. She was not a contributor to these extra volumes, though her sister Ellen is listed.

In 1903 the project was taken on by the Cambridge University Press and work began on the new eleventh edition, designed to be a complete and up-to-date survey of all human knowledge including the rapidly growing sciences. A vital necessity for fulfilling this aim was that contributions should be subscribed speedily and the entire encyclopaedia published at one time. To this end a vast army of experts from all over the world was enlisted, and a team of specialists appointed to advise and guide on the various sciences. The scheme was successfully carried out under the brilliant editorship of Hugh Chisholm,[12] and the new edition appeared in 1910.

The overall adviser for astronomy in the eleventh edition was Agnes Clerke's old friend Simon Newcomb, by this time the doyen of American astronomy. He had last visited the Clerke family in 1898 when over in Britain to receive an honorary degree in Cambridge, and was entertained to lunch and tea by the sisters and their brother.[13] Agnes Clerke herself was a major contributor to the new encyclopaedia, which retained most of the articles which she had written for the earlier, ninth, edition. The main article on astronomy was by Newcomb himself and the history of astronomy – a very long essay – was by Agnes Clerke. (This was the entry which, had she lived, she intended to rectify in order to give Hale his due credit for the invention of the spectroheliograph.) Her substantial entry on ancient astronomies, entitled 'Zodiac', was retained from the earlier edition. She contributed about thirty biographies, including all but one of those (that on the chemist Lavoisier) which she had written for the ninth edition, and was responsible for a revision of the biographies of William and John Herschel, which were originally written by Charles Pritchard.[14]

218 Last days and retrospect

It was while she was engaged on the *Encyclopaedia Britannica* work that Agnes Clerke lost her sister. Yet the work had to go on. She wrote to Newcomb on March 14:

> I know you will be grieved to hear that my sister died of congestion of the lungs on 2nd March. We are very desolate, my poor brother and I, yet not hopeless; for we take comfort in the expectation of a happier life to come. You see, I am writing to you as a real friend such as you have always shown yourself to us, and I feel sure also of Mrs Newcomb's kind sympathy. After a short time I hope to proceed with my work and complete what I have been engaged to do for the *Encyclopaedia Britannica*. I know that 'Algol' has been entrusted to me, and have all the requisite materials at hand.[15]

Chisholm, the encyclopaedia's editor, requested an article of a specified length on 'Nebula' from Agnes Clerke, to be ready by July 1907 after which she expected to be asked to write the article on 'Star'. The articles on Algol and on the history of astronomy were already submitted and most of the biographies were ready. She decided that as there was plenty of time, the nebula article could be deferred for a while.[16] Sadly, neither article was ever written. After Agnes Clerke's death in 1907 they were given to Arthur Stanley (later Sir Arthur) Eddington, then Chief Assistant at the Royal Observatory Greenwich. On both subjects, Eddington refers the reader to Agnes Clerke's *The System of the Stars*. In the article on Star he adds her *Problems in Astrophysics* and Simon Newcomb's *The Stars, A Study of the Universe* and refers the reader to *The System of the Stars* for full references to original papers. Indeed, Eddington's articles are very much what Agnes Clerke would have written herself – except that, in the article on nebula, he makes no rash pronouncement on the 'construction of the heavens', which Agnes Clerke would have confidently described as a one-island universe.

Newcomb died in 1909. It was tragic that neither he nor Agnes Clerke lived to see the monumental eleventh edition of *Encyclopaedia Britannica*, acclaimed by many as its greatest edition ever, in print. Many of Agnes Clerke's major articles and biographies, including that on Galileo, survived in later editions, some in shortened form. She is still listed as a contributor in the sixteenth edition of 1961.

At the time of Ellen's death Agnes was also working on an essay on Descartes for the *Edinburgh Review*, which appeared just half a year before her own death. It was based on a set of books in French, German and English on the life and work of the philosopher, including a centenary edition of his complete works.[17] The *Encyclopaedia Britannica* article on Descartes recommended this essay as a useful sketch of the various available biographies. Though his system of thought was 'irreconcilable with any Christian philosophy', Agnes Clerke gave a very sympathetic account of Descartes the man, comparing and contrasting him with Bacon about whom she had written on an earlier occasion.

Gill's retirement

In the melancholy months after her sister's death, one may well imagine that Agnes Clerke was consoled by the arrival in London in October 1906 of her very dear old friends, David Gill and his wife Bella. Over the years their friendship had never wavered: indeed if anything it had deepened. Gill, too, had been through a busy and stressful time. He had spent many months organising the meeting of the British Association for the Advancement of Science which met at the Cape in 1905. The occasion, as he told Agnes Clerke,[18] was saddened for him by the death of one of his oldest friends, Admiral Wharton, whom he had known since his first astronomical adventure, on the Transit of Venus expedition to Mauritius in 1874. 'His death takes away all the satisfaction and pleasure which we would otherwise have had in the complete success of the BAAS meeting.' This was Gill's last letter to Agnes Clerke from the Cape: his wife's delicate constitution – 'she is more to me than all the earth'[19] – combined with a feeling that his own energy was waning – 'I have not the "go" I used to have'- led him to ask for early retirement a year in advance of the normal age of 65.[20] The news reached London, and Agnes Clerke wrote in her very last letter:

> I am truly glad to have heard from you at last. Thank you with all my heart. I have been longing to know something about you both after those long months of successful labour, followed alas by anxiety and sorrow. Out of consideration for your manifold engagements and the

constant pressure you were under I refrained from writing enquiries which it would have been an additional tax upon you to answer . . . We were prepared for the announcement that you are preparing to leave South Africa. May your homecoming be happy and blessed![21]

Gill retired formally from his post in 1907 and with his wife returned to England in October 1906 to make their final home in London.

Death of Agnes Clerke, 20 January 1907

Agnes Clerke died unexpectedly only ten months after her sister, and in almost identical circumstances: influenza, followed in a few days by pneumonia.

Lady Huggins, who is likely to have visited Agnes Clerke in her last illness, has left a moving account of her death in a letter to George Ellery Hale:

> Before these letters reach you I think it probable that you will have heard the sad news of the death of my dear Friend Agnes Clerke.
>
> Yes, she has gone from us after about a month's illness. It began with just a bad cold which I fear she did not treat early enough. And so, pneumonia and other troubles set in. We hoped to the last, & all that love and skill could do to save her precious life, was done.
>
> To me the blow is very heavy, for besides the fellowship in work, there was also that of a close personal Friendship. I feel very sad and very desolate without her. She was conscious almost to the last, and went to the God she had loved and served, with a sweet and perfect trust and faith. You will be greatly grieved I know; & I wished to give you these few particulars myself.[22]

Agnes Clerke was buried on 23 January next to her parents and sister in Brompton Cemetery after a Requiem Mass in the Catholic Church of the Servites (a religious Order) in Fulham Road.

According to her wishes, expressed in mutual Wills drawn up after their sister's death by her and her brother, her worldly wealth (apart from a few bequests to cousins) was to go to specified Catholic charities in the Diocese of Westminster. These included a fund for the education of priests and substantial bequests to the Mill Hill College for foreign missions, various convents and charitable institutions for the poor including the Catholic Prisoners' Aid Society and Women Prisoners. These bequests show that, like her sister, she was active in Cardinal Vaughan's Crusade of Rescue which operated among deprived people in London. One branch of its activity involved women volunteers working with poor children and orphans. She had a hand in writing a little play about astronomy, 'Stars without Stripes', for convent schoolchildren[23] and perhaps in other charitable ventures not recorded. She was a devout and unostentatious Catholic, ecumenically minded and sensitive to the position of others.

Agnes Clerke had toiled to the last. Two articles of hers appeared after her death, 'Old and new alchemy' in the *Edinburgh Review*, and the account of a paper by Hale, Adams and Gale on sunspot spectra in the *Observatory*.[24] The proof of the latter was returned to the magazine a few days before her death with a note to say that she was lying seriously ill.

The death was announced at the next meeting of the Royal Astronomical Society, where Gill, not long back from South Africa, was among those present. Gill was also present, being inducted as a newly elected member, at the meeting of the Royal Institution on 4 February when the same announcement was made and a vote of condolence to the family recorded.

Tributes

Tributes to Agnes Clerke were numerous. There were first of all the personal ones. Simon Newcomb, on hearing the news, wrote to Gill: 'I was much grieved to hear of Miss Clerke's death following so closely on that of her sister. In past years one of the pleasantest features of my visits to London was the warm and almost affectionate reception by my lady friends at Redcliffe Square.'[25] 'I feel the loss personally very greatly', wrote E.E.Barnard, 'It is a pleasure to remember her kindly greeting to me when I first visited England. At that time her dear mother and sister were alive – but all three have now passed away. There was a heartiness and kindness about the entire family, including the brother, which I have seldom met with elsewhere.'[26] American astronomers were always welcome in the Clerke home whenever they visited London. 'There are dozens of American astronomers who will cordially concur in Professor Barnard's sentiments', wrote T.J.J. See in a published appreciation, 'and among those none joins more heartily than the present writer who also shared the honor of the genial hospitality of Miss Clerke's home while her mother and sister were alive . . . One remembers especially her great simplicity of manner and simple devotion to truth, to which her whole life was consecrated.'[27]

Formal obituary notices extolled her capacity for co-ordinating information and her style of writing. The *Times* obituary notice,[28] which was copied in *Publications of the Astronomical Society of the Pacific* and paraphrased in the Irish *Munster Advertiser*,[29] encapsulates the opinion of her contemporaries:

> . . . her keen insight into the true significance of observed physical facts was as wonderful as her fluency and command of language, so that both from the literary and scientific standpoints she must be ranked as a great scientific writer. No one writing a history of modern astronomy can fail to acknowledge the great debt owed to the masterly array of facts in her 'History'. No worker in the vast field of modern sidereal astronomy opened by the genius of Herschel and greatly widened by the application of the spectroscope to the chemical and physical problems of the universe lacked due recognition by Miss Clerke, who performed as it seemed no other writer could have done the work of collation and interpretation of this enormous mass of material, ever pointing the way to new fields of investigation, often by one pregnant suggestion sweeping aside a whole sheaf of tentative conjectures and indicating, if not the true line – for in many cases the truth is yet to seek – at least a plausible and scientific line worth pursuing.

Nature, in a brief unsigned obituary, called her 'the gifted author of several well-known works' and described her three major books as of 'outstanding merit'. 'Her command of language and acquaintance with astronomical literature were extraordinary, and empowered her to produce books distinguished by literary finish as well as by scientific value.' Agnes would have rejoiced to find the unhappy breach with

Nature thus smoothed over. An indication that the author of the notice was none other than her erstwhile critic Richard Gregory may be gathered from the gentler version of his former denunciation, that 'it is impossible not to feel regret that an enthusiasm so great should have lacked the advantage of laboratory training, which would have enabled Miss Clerke to estimate the real value of the various researches so successfully recorded'.

The *Observatory* mourned 'a powerful writer who was not only able to collate and compile facts but was able to discuss and discriminate'; her place as a long-serving contributor to that magazine would not be easily filled.[30] The gap created by her death was alluded to by others as well. It was hoped that some astronomer would carry on the task of keeping her *History* up to date as she herself had done in successive editions.[31] No-one, in fact, was to make that attempt.

Agnes Clerke was held in particular affection by amateur astronomers. The obituarist in the popular journal *Knowledge* declared that 'as the hand-maiden of astronomy, as one who held a lamp aloft that others might examine its discoveries and its theories, she stood unrivalled in her day'. He called her writing 'a model of style, elegant without affectation, fastidious without sacrifice of meaning, inevitably right in the choice of words'.[32] Altogether, forty-three notices of her death in newspapers and periodicals were collected by her brother.[33]

The longest and most personal account of Agnes Clerke's life was written by one well placed to do it, Margaret Huggins. It was published in the *Astrophysical Journal* and was later expanded into a memoir of the two Clerke sisters to which their brother Aubrey added his own account of their early life in Ireland.[34] He described Lady Huggins' work as 'a labour of love for those whose simple life she records'.

The diligent recorder

At the time of her death Agnes Clerke was a figure of considerable significance in the astronomical world. 'For twenty years she had been to modern astronomy an admirable historian, and had kept before working astronomers clear charts, so to speak, of what was being done, and of what should and might be done', wrote Lady Huggins.[35] She achieved this by working, as she herself wrote, 'without stint of care or pains'.

Agnes Clerke's active life spanned an era of transition, from the days when astronomers like the Hugginses and Lockyer without formal qualifications, working in their private observatories, could be at the forefront of science, to the age of university-trained astrophysicists carrying out their researches in publicly funded institutions. The change had come about more quickly in the United States than it did in Britain, and it was Agnes Clerke more than any other commentator who realised this. An ardent admirer of American astronomers and American methods, she saw that the future lay in large instruments in good observing sites.

On the interpretative side, Agnes Clerke in her *System of the Stars* was an important propagator of the one-galaxy universe, which held sway for decades into the twentieth century. Had she lived longer, she might not have been able to cope with the new theoretical physics. As it was, she was the right person at the right time: someone who had the talent and perseverance to keep in touch with progress, was uninfluenced by worldly ambition, and blessed with a sympathetic personality. She was also lucky in meeting, at crucial moments in her career, many generous advisers – Henry Reeve, Norman Lockyer, David Gill, the Hugginses and George Ellery Hale.

The historian

But however much astronomy has progressed since her death, Agnes Clerke's numerous contributions in the field of the history of astronomy, thoroughly and thoughtfully researched as they were, have endured remarkably well. There is hardly a British or Irish astronomer of note up to the beginning of the twentieth century whose life she has not summarised in the *Dictionary of National Biography*. She is listed as a contributor to all editions of the *Encyclopaedia Britannica* until 1961: her essays on Galileo (1879) (her first signed piece of work) and on Alexander von Humboldt (1881) in the ninth edition of the

Encyclopaedia Britannica survive with only a minimum of revision in the last of these editions, which also carry shortened versions of about a dozen of her biographies of astronomers. Her essay on the history of astronomy, which first appeared in the eleventh (1910) edition, was retained, unaltered, in successive editions (A.S. Eddington being the overall adviser for astronomy) for 40 years, and in later editions the author of the entry on that subject (Angus Armitage) quoted *A Popular History of Astronomy during the Nineteenth Century* as one of his sources.

A Popular History of Astronomy during the Nineteenth Century has in fact never lost its usefulness. Arthur Berry's *Short History of Astronomy* (1898)[36] includes her books in his list of references. Almost a century later, the translator of a German language history of astronomy from 1780 to 1930 states that 'since the appearance of the fourth edition of *A Popular History of Astronomy in the Nineteenth Century* by Agnes M. Clerke there has been no other major English language monograph covering that era of the history of astronomy'.[37] Helen Wright, the distinguished historian of astronomy, in her biography of George Ellery Hale calls her 'the brilliant historian of science'.[38] To Stanley Jaki, philosopher of science and author of a monograph on the history of research on the Milky Way (though he does not always agree with her arguments) she is 'the renowned historiographer of nineteenth century astronomy'.[39] The modern authoritative *General History of Astronomy* (1984)[40] contains a number of references to Agnes Clerke's books and includes her *History* and *Problems in Astrophysics* in a 'further reading' list; it also reproduces historic astronomical photographs, not easily found elsewhere, taken from her books. Michael Crowe, historian and philosopher of science, includes two entire chapters from *The System of the Star*s in his *Modern Theories of the Universe from Herschel to Hubble* (1994),[41] a textbook for graduate liberal arts studies based on primary writings. The most recent history of astronomy, the *Cambridge Illustrated History of Astronomy* (1998),[42] compiled by leading scholars, cites both the *History* and *The System of the Stars*. We may quote here the view of the editor of that important volume, the historian of astronomy, Michael Hoskin:

Of all the works on the history of astronomy ever written, those by Agnes Clerke must surely be the most enduring in value. The successive editions of her *A Popular History of Astronomy during the Nineteenth Century* are goldmines of detailed and reliable information on almost every aspect of astronomy. We know how she did it: her copy of the fourth edition, the last to appear, is interleaved with blank pages, and on these she has begun in meticulous hand to record new discoveries as they appeared in scientific journals, ready to be incorporated in the fifth edition that she did not live to complete. She was a chronicler of her own times, widely respected by astronomers who thought time well spent on letters alerting her to the progress of their science.

Her limitation is encapsulated in the word 'progress'. The introductory chapter deals with the *progress* of the science during the eighteenth century and the *rapid advance* in the nineteenth, and it is followed by Part 1 whose title begins with the word 'progress'. She lived before Einstein, when scientists had little doubt that they were discovering more and more established truths about the universe. She was also dealing with the work of contemporaries or near-contemporaries, whose assumptions and goals she shared. Today's historians see it as their duty to enter into the 'cosmovision' of the scientists they are studying, and to see the world through their eyes and in their terms. without concern for who got it right.

This said, her book on the Herschels and her long encyclopaedia article on Galileo are both remarkable achievements, in which she enters sympathetically into earlier periods. Although during the twentieth century, detailed study of manuscript sources have added greatly to our knowledge of these earlier giants, Clerke's biographical sketches remain reliable outlines that could be recommended to students of today.[43]

Agnes Clerke's enthusiastic style of writing has also been an important element in her continuing popularity. The eminent theoretician Meghnad Saha,[44] was drawn to astronomy and to the problem of the temperatures of the stars by reading her books. Cecilia Payne-Gaposchkin, too, mentions Agnes Clerke among those predecessors who had thrilled at the sight of beautiful stellar spectra.[45]

In the preface to the first edition of her *History* (1885), Agnes Clerke laid down her aim 'to give to each individual discoverer, strictly and impartially, his due'. 'Impartiality is one of the highest virtues of a historian', said one reviewer of the *History*,[46] 'and it must be allowed that Miss Clerke has shown this to a high degree. Indeed, if we may permit ourselves to criticize what is on the whole an admirable performance, we should say that she sometimes carries the virtue of impartiality to excess'. Her obituarist and friend Lady Huggins made the same comment in more elegant terms. 'She was so incapable of meanness that she even incurred danger as a historian in crediting too readily all workers with her own high ideals.'[47] To Margaret Huggins (ever watchful of challenges to her husband's pre-eminence) some astronomers, like the denizens of *Animal Farm*, 'are more equal than others'. However, this over-conscientious attitude on the part of Agnes Clerke constitutes a positive advantage to any present-day historian of astronomy who may wish to trace the activities of the lesser players on the nineteenth-century stage: there are well over 500 names listed in the index of the last edition of the *History*, and a like number of dates in her chronological Table, from '1774 March 4. Herschel's first observation' to '1901 September 19. Unveiling of the "Victoria" telescope at the Royal Observatory, Cape of Good Hope'.

By the end of her life Agnes Clerke was well aware of the pitfalls of writing the history of a subject at close range. When she launched the first edition back in 1885, astrophysics, or the 'new astronomy' as it was then called, seemed easy compared with the old classical astronomy. It was descriptive and mathematics-free, 'more popular in its nature . . . more easily intelligible . . . less remote from ordinary experience'. But rapid changes had taken place since then. She remarked in 1902 how her earlier editions now had 'a superannuated look'. She went on:

> The writing of history is a strongly selective operation, the outcome being valuable just in so far as the choice to reject and what to include has been judicious; and the task is no light one of discriminating between barren speculation and ideas pregnant with coming truth. To the possession of such prescience of the future as would be needed to do this effectually, I can lay no claim; but diligence and sobriety of thought are ordinarily within reach, and these I have exercised to good

purpose if I have succeeded in rendering the fourth edition of *A Popular History of Astronomy during the Nineteenth Century* not wholly unworthy of a place in the scientific literature of the twentieth century.[48]

At the opening of the twenty-first century, we can agree that she succeeded in her task.

17 Epilogue

Of Agnes Clerke's London-based friends, Sir David Gill continued to take an active interest in astronomy after he retired. He was awarded the gold medal of the Royal Astronomical Society in 1908 and was the Society's President in 1909–11. He died in 1914; Lady Gill survived him until 1919.

The Tulse Hill observatory ceased to function in 1908, when Sir William Huggins was 85. The instruments reverted to the Royal Society, which had originally supplied them, and were given to the Cambridge University Observatory. The Hugginses in their retirement collected and edited their scientific papers, which were published in a handsome volume, *The Collected Scientific Papers of Sir William Huggins*, in 1909. Sir William Huggins died in 1910; Lady Huggins died in 1916.

In a reorganisation of British astronomy, Sir Norman Lockyer's solar observatory at South Kensington was transferred to Cambridge University in 1910, greatly to his disappointment. He then set up a private observatory at Sidmouth, Devon, later named the Norman Lockyer Observatory, which opened in 1913 and still flourishes. He died in 1920.

Walter Maunder retired in 1913 after 40 years of service with the Royal Observatory, but resumed his duties after the outbreak of the First World War, when many of the Greenwich staff were absent on military service. His wife Annie also returned to Greenwich from 1915 to 1920, as a volunteer. Both continued to be active in the British Astronomical Association until their deaths. Walter died in 1928, Annie in 1947.

After Agnes Clerke's death her brother Aubrey lived on alone in Redcliffe Square for a further 16 years. He never married. He continued

Figure 17.2 Plaque honouring Agnes and Ellen Clerke on the house where they were born.

to work at his Barrister's chambers in Lincoln Inns Fields until his death, a familiar sight on his bicycle but something of a recluse in his private life. Aubrey Clerke died on 23 December 1923, aged 80, and is buried with his parents and sisters in Brompton Cemetery.

Aubrey's nearest relative was his first cousin H.H.P. Deasy, the explorer. The books that had originated in the Clerke library in Skibbereen were taken by him to his home in Co. Tipperary, where they are preserved with the Deasy family papers. His son, the late Rickard J.G. Deasy, who was Aubrey Clerke's godson, took a great interest in the life of Agnes Clerke, an interest enthusiastically maintained by his son, Rickard H. Deasy.

Figure 17.1 (opposite) Crater Clerke on the Moon. Antonin Rükl, map of the moon, courtesy Kluwer Academic Publishers.

Agnes Clerke's contributions to astronomy have not gone unhonoured. She is commemorated in a small crater on the Moon, one of the additional features revealed on the Lunar Orbiter photographs, named by NASA in 1981.[1] 'Clerke', 7 km in diameter at latitude and longitude 22 °N, 30 °E, lies on the border of the Sea of Serenity close to the spot where Apollo 12 landed on 11 December 1972. (Figure 17.1) On 11 July 1999 a plaque with portraits of Agnes and Ellen Clerke was unveiled on the house in Skibbereen where the sisters were born and grew up (Figure 17.2), a tribute by the Irish scientific and local communities to a remarkable family and to two devoted scholars.[2]

Notes

ABBREVIATIONS IN THE NOTES

AMC Agnes Mary Clerke

Appreciation Lady Huggins Hon. Mem. R.A.S. (1907). *Agnes Mary Clerke and Ellen Mary Clerke, an Appreciation.* Printed for private circulation.

British Library The Manuscripts Collection, The British Library, London.

Cal Tech Hale Papers, Institute Archives, California Institute of Technology, Pasedena, California.

Cape/RGO Gill–Clerke correspondence, Cape archives in Royal Greenwich Observatory archives, Cambridge University Library. Files RGO 15/126–130.

DNB *Dictionary of National Biography*, Compact Edition, Oxford University Press 1975.

Exeter University Archives in the University Library, University of Exeter.

GF George Forbes, *David Gill, Man and Astronomer*, John Murray London 1916.

HCO Pickering correspondence, Library, Harvard-Smithsonian Center for Astrophysics, Cambridge, Mass.

Heidelberg Wolf correpondence (Heid. Hs. 3695E), Universtätsbibiothek, University of Heidelberg.

Library of Congress Newcomb correspondence, Manuscript Division, The Library of Congress, Washington.

MLS Mary Lea Shane archives of the Lick Observatory, University of California, Santa Cruz.

RGO Royal Greenwich Observatory. The RGO archives at Cambridge University Library are quoted by permission of the syndics of Cambridge University Library.

Royal Institution Archives of the Royal Institution (RI.8 and 10), London.

UW University of Wisconsin-Madison Division of Archives, B134 Memorial
Library, Madison.

Yerkes Archives, Yerkes Observatory, The University of Chicago, Williams Bay Wi.

CHAPTER I. FAMILY BACKGROUND IN COUNTY CORK

1 M.T. Brück (1993). Ellen and Agnes Clerke, scholars and writers. *Seanchas
   Chairbre*, 3, 23–42.
2 Obituary. *Monthly Notices of the Royal Astronomical Society*, 10, 78, 1849–50;
   DNB.
3 Obituary. *Skibbereen Eagle*, 4 May 1867.
4 *Southern Reporter*, 4 December 1824.
5 *Burke's Irish Family Records*. 1976 London: Burke's Peerage Ltd.
6 Thomas D. Clareson (1985). Fitz-James O'Brien. In E.F.Bleiler (ed.) *Supernatural
   Fiction Writers*. New York: Charles Scribner's sons. I thank Rickard H.
   Deasy for this reference and other information on the family.
7 According to Agnes Clerke's brother Aubrey, their mother believed herself to be
   a cousin of the Earl of Fermoy. The Earl of Fermoy, of the Irish Peerage,
   formerly Edward Burke Roche, whom Rickard Deasy succeeded as Liberal
   member of Parliament for County Cork, was related through his wife to
   Burke. (Aubrey Clerke's letters to his cousin, Deasy family papers.)
8 Copies of letters to his son Rickard Morgan Deasy. Deasy family papers.
9 Michael O'Connell (1982). The craft of the cooper in Clonakilty. *Seanchas
   Chairbre*, 1, 44–8.
10 Tomas Tuipear. *Historical Walk of Clonakilty and its Sea-front*. Clonakilty:
   Local History Society.
11 Rev. James Coombes (1978). *Southern Star*, 23 September.
12 Decie and his descendents are listed in *Burke's Irish Family Records*.
13 Mrs Deasy, letter to her son Rickard, c. 1845. Deasy family papers.
14 Ellen Deasy is not listed in *Burke's Irish Family Records* but she is included in a
   family tree constructed by Aubrey Clerke and in the Ursuline Convent list
   of pupils.
15 Ursuline Convent, Cork, archives.
16 Information from Rickard H. Deasy, who believes that Sheahan's tutoring
   accounts to a great extent for Catherine Clerke's unusually high level of
   education, imparted to her daughters.
17 Ursuline Convent, Cork, archives.
18 Prospectus of the Ursuline Convent (n.d., pre 1850). Maria Luddy (1995).
   Women in Ireland 1800–1918, A Documentary History, p. 106. Cork:
   University Press.

[19] W.M. Thackeray (1900: Centenary edition). Chapter 6: Cork – the Ursuline Convent. In *The Irish Sketch Book*, London, Edinburgh and New York: Thomas Nelson and Sons.

[20] Letter of Aubrey Clerke to H.H.P. Deasy, 6 September 1920. Deasy family papers.

[21] Pat Cleary (1983). *Seanchas Chairbre*, 2, 40–65.

[22] The British Association for the Advancement of Science debated a Report on the Skibbereen calamity in its Statistics Section at its meeting in 1850. *BAAS Report* 1850, 149–50.

[23] Cecil Woodham-Smith (1962). *The Great Hunger 1845–49*, p. 199. London: Hamish Hamilton.

[24] Pat Cleary. Unpublished communication to the author.

[25] AMC (1886). Aurora Borealis, *Edinburgh Review*, 74, 416–17.

[26] Joseph Lee (1989). *The Modernisation of Irish Society 1848–1918*, pp. 57–58. Dublin: Gill and Macmillan. O'Donovan Rossa's one-time shop is now suitably commemorated by a handsome plaque.

[27] Aubrey Clerke (1907). Preface to *Appreciation*.

[28] Notice of death of J.W. Clerke. *Eagle and County Cork Advertiser*, 8 March 1890.

[29] Ursuline Convent, Cork, archives.

[30] Quoted in C. Clear (1987). *Nuns in 19th Century Ireland.* Dublin: Gill and Macmillan; Washington: Catholic University of America Press.

[31] Ellen M. Clerke (1888). Irish industries. *Dublin Review*, 21, 378–99.

[32] James Coombes (1982). Catherine Donovan (1788–1858), Educational Pioneer. *Seanchas Chairbre*, 1, 51–8.

CHAPTER 2. IRELAND AND ITALY

[1] Aubrey Clerke (1907). Preface to *Appreciation*.

[2] AMC (1903). Magazine interview. Copy in the Deasy family papers.

[3] Deasy family library.

[4] Jeremiah Joyce (1763–1813) was a popular writer on science. *DNB*.

[5] AMC. Entry on Nichol, *DNB*.

[6] Mary Somerville (1824). *The Connexion of the Physical Sciences.* London: John Murray.

[7] J.W. Cross (1885). *George Eliot's Life as Related in her Letters and Journals*, Volume 1, p. 65. New York: Harper and Brothers.

[8] Hector Macpherson (1903). *Astronomers of Today and their Work*, chapter on Agnes Clerke. London and Edinburgh: Gall and Inglis.

[9] AMC. Entry on John Herschel. *DNB*.

[10] *DNB*.

[11] Christopher Brooke (1985). *A History of Gonville and Caius College*, p. 195. Woodbridge: The Boydell Press; *DNB*.

[12] The volumes, all inscribed 'Aubrey Clerke, 19 August 1859', are in the Deasy library.

[13] The land is mentioned in the notice of J.W. Clerke's death in the local paper; its sale, then in progress, is noted in Aubrey's Will dated 1914.

[14] Obituaries. *Freeman's Journal*, 7 May 1883; *Times* (London), 7 and 8 May 1883; *DNB*.

[15] James Joyce (1914; reprint 1956). *Dubliners*. London: Penguin Books.

[16] Peter Costello (1992). *James Joyce, The Years of Growth (1882–1915)*. London: Kyle Cathie Ltd.

[17] Lady Huggins (1907). *Appreciation*.

[18] Robert Sidney Pratten. *DNB*.

[19] Royal Irish Academy archives.

[20] G.F. Mitchell (1985). Antiquities. In T. O'Raiftearaigh (ed.) *The Royal Irish Academy, A Bicentennial History 1785–1985*, p. 102. Dublin: Royal Irish Academy.

[21] G.F. Mitchell. op. cit., p. 128.

[22] Robert Welch (ed.) (1966). *The Oxford Companion to Irish Literature*, p. 599. Oxford: Clarendon Press.

[23] AMC (1886). On the Aurora Borealis. *Edinburgh Review*, **74**, 416–47.

[24] Ellen M. Clerke (1899). *Fable and Song in Italy*. London 1899.

[25] On the Irish missionary saints on the Continent (1896) and the poet of the Young Ireland Movement, Denis Florence McCarthy (1892), in *Dublin Review*.

[26] W. Valentine Ball (1915). *Reminiscences and Letters of Sir Robert Ball*. London: Cassell and Co.

[27] Ellen M. Clerke. Six articles on Italy published in the *Cornhill Magazine*, 1879–81; also one on rural life in Italy in *Edinburgh Review*, January 1887.

[28] Ellen M. Clerke (1902). *Flowers of Fire*. London: Hutchison and Co.

[29] AMC to Edward Holden, 20 September 1890, in response to a photograph which he sent her of forest fires near Lick Observatory in California. MLS.

[30] Katherine Frank (1995). *Lucie Duff-Gordon, A Passage to Egypt*. London: Penguin.

[31] Interestingly, the next occupant of that address after the Clerkes moved to a larger house was the newly married Harry Trollope, son of the novelist Anthony Trollope. (Victoria Glendinning (1992). *Trollope*, p. 504. London: Hutchinson). This raises the possibility that the Clerkes may have known the Trollope family. Anthony Trollope was employed by the Post Office in

Ireland from 1841 to 1859 in the course of which he travelled all over the south and west of Ireland, while his brother Thomas, an expatriate in Florence, was still living there during the Clerke sisters' time.

CHAPTER 3. LONDON, THE LITERARY SCENE

[1] AMC. Article on Henry Reeve. *DNB*.

[2] J.K. Laughton (1898). *Memoirs of the Life and Correspondence of Henry Reeve CB DCL*, vol. 2. London: Longman.

[3] Walter Edward Houghton (ed.) (1972). *The Wellesley Index to Victorian Periodicals 1824–1900*, vol. 2, Toronto: University of Toronto Press.

[4] Dr Russell had no personal connection with either of the Clerkes in their Dublin years, or with Judge Deasy. Deasy does not appear in the biography of Russell (Ambrose Macauley (1983). *Dr Russell of Maynooth*, London: Darton Longman and Todd) nor in the Maynooth archives. I thank Father Corish, historian and archivist at Maynooth, for this information.

[5] Robert O'Neil (1995). *Cardinal Herbert Vaughan, Archbishop of Westminster, Bishop of Salford, Founder of the Mill Hill Missionaries*. Tunbridge Wells: Burns and Oates.

[6] Mary R.S. Creese (1998). *Ladies in the Laboratory? American and British Women in Science, 1800–1900: A Survey of their Contributions to Research*, pp. 324–5. Lanham and London: The Scarecrow Press.

[7] Robert O'Neil. op. cit., p. 451.

[8] *Dublin Review*, **118**, p. 245, 1896.

[9] *Athenaeum*, **60**, 330, 1882.

[10] E.C. Patterson (1983). *Mary Somerville and the Cultivation of Science 1815–1840*. The Hague: Nijhoff; M.T. Brück (1996). Mary Somerville, mathematician and astronomer of underused talents. *Journal of the British Astronomical Association*, **106**(4), 201–6.

CHAPTER 4. THE HISTORY OF ASTRONOMY

[1] Robert Grant (1852). *History of Physical Astronomy from the Earliest Ages to the Middle of the Nineteenth Century*. London: Henry G. Bohn.

[2] AMC (1880). The chemistry of the stars. *Edinburgh Review*, **152**, 408–43.

[3] AMC to Holden. 15 January 1884. UW.

[4] Ellen M. Clerke (1881). *The Flying Dutchman and other Poems*. London: W. Satchell & Co.

[5] Agnes M. Clerke (1885). *A Popular History of Astronomy during the Nineteenth Century*. Edinburgh: Adam and Charles Black.

[6] [R.J. Mann] (1886). *Edinburgh Review*, **163**, 373–405.

[7] *Dublin Review*, **15**, p. 219, 1886.

[8]   R.S. Ball (1886). *Nature*, **33**, 313.
[9]   Robert S. Ball (1886). *The Story of the Heavens*. London: Cassell and Company.
[10] E.W. Maunder (1886). *Observatory*, **98**, 126.
[11] Lady Huggins, *Appreciation*.
[12] *Knowledge*, **7** (no. 1 New Series), 1885–86, 69.

CHAPTER 5. A CIRCLE OF ASTRONOMERS

[1]   Helen Wright (1987). *James Lick's Monument*, p. 45. Cambridge: Cambridge University Press.
[2]   D.E. Osterbrock (1984). The rise and fall of Edward Holden, Part 1. *Journal for the History of Astronomy*, **15**, 81–127.
[3]   Carolyn Heilbrun (1961). *The Garnett Family*. London: George Allan and Unwin.
[4]   AMC to Holden. 15 January 1884. UW.
[5]   AMC to Holden. [January 1884]. UW.
[6]   D.E. Osterbrock. op. cit,. p. 84.
[7]   AMC to Holden. 31 January 1886. MLS.
[8]   AMC to Holden. 25 July 1886. MLS.
[9]   AMC to Holden. 23 November 1886. MLS.
[10] *Publications of the Astronomical Society of the Pacific*, **1** No 2, 1889.
[11] *Publications of the Astronomical Society of the Pacific*, **1** No 3, 1889.
[12] D.E. Osterbrock. op. cit.
[13] The Great Lick Telescope. *Knowledge*, **10** (new series 2). July 1887, pp. 205–7.
[14] idem, pp. 209–10.
[15] H.A. Brück (1992). Lord Crawford's observatory at Dun Echt 1872–1892. *Vistas in Astronomy*, **35**, 81–138.
[16] Copeland to AMC. 1884. Dun Echt archives in the library of the Royal Observatory Edinburgh.
[17] A. Fowler. Appendix to T. Mary Lockyer and Winifred L. Lockyer (1928). *Life and Work of Sir Norman Lockyer*, p. 249. London: Macmillan.
[18] T. Mary Lockyer and Winifred L. Lockyer (1928). *Life and Work of Sir Norman Lockyer*, p. 117. London: Macmillan.
[19] AMC to Holden. 23 November 1886. MLS.
[20] AMC to Lockyer. 10 July 1886. Exeter University Archives.
[21] ibid.
[22] AMC to Lockyer [1887?]. Exeter University Archives.
[23] AMC to Lockyer. 6 May 1887. Exeter University Archives.
[24] Notes. *Nature*, **36**, p. 42, May 1887.
[25] AMC (1883). Old observations of Saturn. *Observatory*, **6**, 341.
[26] Lady Huggins. *Appreciation*.

[27] AMC to Pickering. 26 October 1886. HCO.

[28] AMC to Pickering. 10 June 1887. HCO.

[29] ibid.

[30] *Edinburgh Review*, **78**, 23–46, 1888.

CHAPTER 6. A VISIT TO SOUTH AFRICA

[1] Agnes M. Clerke (1889). A southern observatory. *The Contemporary Review*, **4**, 380–92. (Reprinted in *Annual Report of the Board of Regents of the Smithsonian Institution for 1891*. Washington, 1893).

[2] H.A. Brück (1992). Lord Crawford's observatory at Dun Echt 1872–1892. *Vistas in Astronomy*, **35**, 81–138.

[3] [AMC] (1888). *Edinburgh Review*, **78**, 23–46.

[4] George Forbes (1916). *David Gill, Man and Astronomer*, p. 195. London: John Murray.

[5] Rickard Deasy (1999). *Southern Star*, Skibbereen, 17 July, quoting a magazine interview with AMC.

[6] Gill to AMC. 6 December 1887. GF.

[7] Gill to AMC. 15 April 1888. GF.

[8] AMC to Holden. 27 May 1888. MLS.

[9] Gill to AMC. 16 June 1888. GF.

[10] Gill to AMC. 28 November 1888. GF.

[11] A.M. Clerke (1888). Southern star spectra. *Observatory*, **11**, 429–32; (1889) Some southern red stars. *Observatory*, **12**, 134–6.

[12] D.J.K. O'Connell (1956). Photographic light curve of Eta Carinae. *Vistas in Astronomy*, **2**, 1165–71.

[13] Gill to Ellen Clerke. 18 September 1888. GF.

[14] AMC. *The Contemporary Review*, op. cit.

[15] AMC (1889). The Cape in 1888. *Dublin Review*, **21**, 106–13.

[16] AMC. *The Contemporary Review*, op. cit.

[17] Gill to W. Elkin. 6 November 1888. GF.

[18] Gill to AMC. 15 April 1888. GF.

[19] AMC to Gill. 20 June 1892. Cape/RGO.

[20] AMC to Gill. 1 December 1889. Cape/RGO.

[21] Derek Howse (1975). *Greenwich Observatory: Volume 3, The Buildings and Instruments*. London: Taylor and Francis; Allan Chapman (1989). William Lassell (1799–1880), Practitioner, patron and 'Grand Amateur' of Victorian astronomy. *Vistas in Astronomy*, **32**, 341–70.

[22] 'Salaries of Lady Computers, April 1890'. RGO 7/140.

[23] M.T. Brück (1995). Lady Computers at Greenwich in the early 1890s. *Quarterly Journal of the Royal Astronomical Society*, **36**, 83–95.

24 AMC to Gill. 1 June 1890. Cape/RGO.

25 H.M. Harrison (1994). *Voyager in Time and Space, The Life of John Couch Adams Cambridge Astronomer*, p. 226. Lewes: The Book Guild.

26 Copy enclosed in a letter from Holden to AMC. 6 August 1889. MLS.

27 AMC to Holden. 1 September 1889. MLS.

28 Agnes M. Clerke (1889). Photographic star-gauging, *Nature*, 40, 344–6; reprinted in *Annual Report of the Smithsonian Institution for 1891*. Washington 1893.

29 A.M. Clerke (1889). The spectra of the Orion Nebula and of the Aurora. *Observatory*, 12, 366–70.

30 B.J. Becker (1996). Dispelling the myth of the able assistant: William and Margaret Huggins at work at the Tulse Hill Observatory. In H.M. Pycor *et al.* (eds.) *Creative Couples in the Sciences*. New Brunswick: Rutgers University Press. Dr Becker concludes that the Hugginses were 'complementary collaborative investigative partners' whose image as Ruskin's classic Victorian couple of creator and helper was of their own creation.

31 H.A. and M.T. Brück (1988). *The Peripatetic Astronomer, the Life of Charles Piazzi Smyth*, p. 192. Bristol and Philadelphia: Adam Hilger.

32 W. Valentine Ball (ed.) 1915. *Reminiscences and Letters of Sir Robert Ball*, p. 129. London: Cassell and Company Ltd.

33 Gill to AMC. 15 April 1888. GF.

34 Agnes M. Clerke (1889). *Geschichte der Astronomie während des 19 Jahrhunderts*. Autorisierte deutsche Ausgabe von H. Maser [authorised German translation by H. Maser]. Berlin: Julius Springer.

CHAPTER 7. *THE SYSTEM OF THE STARS*

1 Gill to AMC. 9 March 1889. GF.

2 Gill to AMC. 15 March 1889. GF.

3 Gill to AMC, note 1.

4 Gill to AMC. 1 July 1889. GF.

5 J.B. Hearnshaw (1986). *The Analysis of Starlight*, p. 147. Cambridge: Cambridge University Press.

6 Vogel's Potsdam correspondence is now lost. A letter from Agnes Clerke, believed to be to Vogel, was sold at auction in 1940. Its present whereabouts is not known. (Letter from Dr W. Dick to the author, 10 January 1999.)

7 AMC to Holden. 12 December 1889. MLS.

8 AMC (1888). *Observatory*, 11, 79–84.

9 A.J. Meadowes (1972). *Science and Controversy, A Biography of Sir Norman Lockyer*, Chapter 7. Cambridge, Mass.: The MIT press.

[10] *System of the Stars*, 1st edition, p. 88.

[11] B.J. Becker (1993). *Eclecticism, Opportunism, and the Evolution of a New Research Agenda: William and Margaret Huggins and the Origins of Astrophysics*, Volume 2, p. 281. PhD Dissertation, Johns Hopkins University, Baltimore. It transpired that there was a slight error in the Hugginses' wavelength of the green nebular line, which did not, however, invalidate their conclusion.

[12] AMC (1889). *Observatory*, **12**, 366–70.

[13] AMC to Gill. 1 February 1890. Cape/RGO.

[14] M.L. Huggins (1898). *Astrophysical Journal*, **8**, 54.

[15] AMC to Lockyer. 31 December 1889. Exeter University archives.

[16] J. Norman Lockyer (1894). *The Dawn of Astronomy*. London: Cassell and Co.

[17] AMC to Gill. 12 October 1890. Cape/RGO.

[18] AMC to Holden. 9 November 1889. MLS.

[19] D.E. Osterbrock (1984). *James E. Keeler: Pioneer American Astrophysicist*, p. 95. Cambridge: Cambridge University Press.

[20] AMC to Holden. 24 January 1890. MLS.

[21] AMC to Gill. 1 February 1890. Cape/RGO.

[22] Holden to AMC. 25 July 1889. MSL.

[23] AMC to Holden. 15 August 1889. MLS.

[24] D.E. Osterbrock (1984). *Journal for the History of Astronomy*, **15**, 97.

[25] Holden to AMC. 8 August 1889. MLS.

[26] Barnard to AMC. 17 February 1890. MLS.

[27] AMC to Holden. 23 October 1889. MLS.

[28] Holden to AMC. 2 July 1889. MLS.

[29] Gill to AMC. 18 September 1889. Cape/RGO.

[30] AMC to Gill. 3 December 1889. Cape/RGO.

[31] AMC (1890). Stellar spectroscopy at Lick Observatory. *Observatory*, **13**, 46–9.

[32] Michael Hoskin and Owen Gingerich (1997). The 1900 consensus. In David Dewhirst and Michael Hoskin (ed.). *The Cambridge Illustrated History of Astronomy*, p. 326. Cambridge: Cambridge University Press.

[33] *System of the Stars*, 1st edition, p. 368; 2nd edition, p. 348.

[34] *System of the Stars*, 1st edition, p. 372; 2nd edition, p. 355.

[35] *System of the Stars*, 1st edition, p. 376; 2nd edition, p. 358.

[36] *System of the Stars*, 1st edition, p. 380; 2nd edition, p. 361. Interestingly, however, Agnes Clerke believed in the existence of 'dark nebulae' composed of 'dark' (invisible) stars (*Problems in Astrophysics*. p. 541).

[37] *System of the Stars*, 1st edition, p. 378.

[38] *System of the Stars*, 1st edition, p. 290. This and the previous remark do not appear in the second edition.

39 *System of the Stars*, 1st edition, p. 357; 2nd edition, p. 375.

40 *System of the Stars*, 1st edition, p. 396.

41 *System of the Stars*, 2nd edition, p. 374.

42 AMC to Gill. 1 December 1889. Cape/RGO.

43 AMC to Holden. 10 March 1890. MLS.

44 AMC to Gill. Easter Sunday 1890. Cape/RGO.

45 AMC to Gill. 1 June 1890. Cape/RGO.

46 AMC to Gill. 3 September 1890. Cape/RGO.

47 Alan H. Batten (1988). *Resolute and Undertaking Characters, The Lives of Wilhelm and Otto Struve*, Chapter 16. Dortrecht: Reidel.

48 AMC to Gill. 3 November 1890. Cape/RGO.

49 Holden to AMC. 21 January 1891. MLS.

50 M.L. Huggins (1890). *Observatory*, **13**, 382.

51 F (25 December 1890). *Nature*, **43**, 169.

52 AMC to Holden. 22 February 1891. MLS.

53 G.E. Hale (1891). *Publications of the Astronomical Society of the Pacific*, **3**, 180.

54 Helen Wright (1994). *Explorer of the Universe, A Biography of George Ellery Hale*, p. 71. New York: American Institute of Physics.

55 *The Athenaeum*. 9 May 1891.

56 Quoted in E.C. Patterson (1983). *Mary Somerville and the Cultivation of Science, 1815–1840*, p. 98. The Hague: Martinus Nijhoff Publishers.

CHAPTER 8. SOCIAL LIFE IN SCIENTIFIC CIRCLES

1 Colin Ronan (1990). *Journal of the British Astronomical Association*, **101**, 107–110.

2 Mary Creese (1998). Elizabeth Brown, solar astronomer. *Journal of the British Astronomical Association*, **108** (4), 193–7.

3 The British Astronomical Association (1890). *Observatory*, **13**, 294.

4 AMC to Gill. 12 October 1890. Cape/RGO.

5 Turner to Lockyer. Quoted in A.J. Meadowes (1972). *Science and Controversy, A Biography of Sir Norman Lockyer*, p. 225. Cambridge Mass.: The MIT Press.

6 AMC to Holden. 23 October 1889. MLS.

7 AMC (1890). The System of Zeta Cancri. *Publications of the Astronomical Society of the Pacific*, **2**, 188.

8 R.A. Jarrell (1988). *The Cold Light of Dawn, A History of Canadian Astronomy*. Toronto: University of Toronto Press.

9 AMC (1892). *Observatory*, **15**, 334.

10 AMC to Gill. 1 February 1890. Cape/RGO.

[11] William Huggins (1891). Presidential address. *Report of the British Association for the Advancement of Science,* **61**, 3–37.

[12] AMC to Gill. 20 August 1891. Cape/RGO.

[13] Helen Wright (1994). *Explorer of the Universe, A Biography of George Ellery Hale (History of Modern Physics and Astronomy Volume 14).* New York: American Institute of Physics.

[14] AMC to Gill. 20 August 1891. Cape/RGO.

[15] Helen Wright. op. cit., p. 83.

[16] AMC to Hale. 17 February 1892. Yerkes.

[17] Copeland's diary. Archives of the Royal Observatory Edinburgh.

[18] R.S. Ball (1893). *The Story of the Sun.* London: Cassell and Co.

[19] AMC to Hale. 1 November 1891. Yerkes.

[20] AMC to Hale. 2 January 1893. Yerkes.

[21] AMC to Gill. 16 November 1893. Cape/RGO.

[22] AMC to Gill. 9 September 1894. Cape/RGO.

[23] John A. Brashear (reprint 1988). *A Man who Loved the Stars,* p. 120. Pittsburgh: University of Pittsburgh Press.

[24] M.T. Brück (1995). Lady Computers at Greenwich in the early 1890s. *Quarterly Journal of the Royal Astronomical Society,* **36**, 83–95.

[25] J.L.E. Dreyer (1923). In J.L.E. Dreyer and H.H. Turner (eds.). *History of the Royal Astronomical Society,* Volume 1, p. 234. London: Royal Astronomical Society.

[26] AMC to Gill. 1 November 1893. Cape/RGO.

[27] Hector McPherson (1907). *Popular Astronomy,* **15**, 165.

[28] I am grateful to Mrs Sheila Edwards, Librarian of the Royal Society, for this information on the position of women visitors.

[29] AMC to Gill. 1 June 1890. Cape/RGO.

[30] AMC to Sir Frederick Bramwell. 7 March 1893. Royal Institution archives.

[31] Holden to AMC. 4 April 1893. MLS.

[32] AMC [1903]. Magazine interview, Deasy family papers.

[33] T.J.J. See (1907). Some recollections of Miss Agnes M. Clerke. *Popular Astronomy,* **15**, 323.

[34] AMC to Gill. 12 October 1890. Cape/RGO.

[35] *The Book of Trinity College Dublin 1591–1891.* Belfast 1892. Other astronomers among the guests were Norman Lockyer, Ralph Copeland and Isaac Roberts. Richard Garnett of the British Museum was also there.

[36] AMC to Gill. 11 July 1892. Cape/RGO.

[37] AMC to Gill. 12 October 1890. Cape/RGO.

[38] T.J.J. See. op. cit.

[39] AMC to Wolf. 7 June 1892. Heidelberg.

[40] There are fourteen letters from AMC to Wolf written between 1892 and 1906 in the Heidelberg University archives.

[41] EMC to Holden. 8 October [1893].

[42] T. Nigel Brown (1971). *The History of the Manchester Geographical Society 1884–1950*. Manchester: Manchester University Press.

[43] E.M. Clerke (1891). On the aborigines of western Australia. *Report of the British Astronomical Association 1891*, p. 716.

[44] Mary R.S. Creese (1998). *Ladies in the Laboratory? American and British Women in Science, 1800–1900*, p. 324. Lanham and London: The Scarecrow Press, Inc.

[45] E.M. Clerke (1889). The dock labourers strike: the labour market of East London. *Dublin Review*, 22, 386–406.

[46] *DNB*.

[47] E.M. Clerke (1889). The principles of '89, *Dublin Review*, 22, 118–39.

[48] Michael Walsh (1990). *The Tablet 1840–1990, A Commemorative History*. London: The Tablet Publishing Company.

[49] Obituary of E.M. Clerke. *Tablet*, 10 March 1906.

[50] Lady Huggins in her *Appreciation* states that Ellen was on the staff of the *Tablet* for 20 years.

[51] AMC to Holden. 18 December 1891. MLS.

[52] Michael Walsh. op. cit. p. 23.

[53] Michael Walsh. idem.

[54] E.M. Clerke (1892). *Observatory*, 15, 274.

[55] E.M. Clerke (1892). *Jupiter and its System*. London.

[56] E.S. Holden (1892). *Publications of theAstronomical Society of the Pacific*, 4, 267.

[57] AMC to Hale. 11 October 1892. Yerkes.

[58] EMC to Holden. 25 February [1893]. MLS.

[59] E.M. Clerke (1893). *The Planet Venus*. London.

[60] EMC to Holden. 25 February [1893]. MLS.

CHAPTER 9. HOMER, THE HERSCHELS AND A REVISED *HISTORY*

[1] AMC to Holden. 13 December 1891. MLS.

[2] AMC (1892). *Familiar Studies in Homer*. London: Macmillan.

[3] AMC [1903]. Magazine interview, Deasy family papers.

[4] Schliemann's excavations at Mycenae and Flinders Petrie's excavations in Egypt.

[5] Sara Schnechner Genuth (1992). *Journal for the History of Astronomy*, 23, 293–8. Agnes Clerke's book is not mentioned in this paper, though she is referred to in another connection. Dr Genuth's chief purpose is to counter the claim made by many writers since 1910 that Homer in the *Iliad* referred to comets as auguries of evil.

6   AMC to Samuel Butler. British Library 440434.

7   Henry Festing Jones (1922). Preface to the second edition of Samuel Butler, *The Authoress of the Odyssey*. London: Jonathon Cape.

8   Holden to AMC. 2 November 1892. MLS.

9   AMC. Preface to the third edition, May 1893.

10 AMC (1892). Nova Aurigae. *Observatory*, **15**, 334–9.

11 *Problems in Astrophysics*, p. 378.

12 *Observatory*, **16**, 300, 1893.

13 *Astronomy and Astrophysics*, **14**, 846, 1893.

14 *Nature*, **49**, 2, 1893.

15 A.J. Meadowes (1972). *Science and Controversy, A Biography of Sir Norman Lockyer*, p. 227 and Chapter 8. Cambridge Mass.: MIT Press.

16 W.H. Wesley (1895). *Knowledge*, 1 February, 25–7; Editorial Note, ibid. p. 25.

17 AMC to Gill. 3 October 1894. Cape/RGO.

18 AMC (1893). Proctor's Old and New Astronomy. *Edinburgh Review*, **177**, 544–64.

19 Richard A. Proctor, completed by A. Cowper Ranyard (1892). *Old and New Astronomy*. London: Longman Green and Co.

20 S.L. Jaki (1972). *The Milky Way, An Elusive Road for Science*, p. 271 and references therein. New York: Science History Publications.

21 Michael Hoskin (ed.) (1997). *Cambridge Illustrated History of Astronomy*, p. 321. Cambridge: Cambridge University Press.

22 Wilson's diary, unpublished.

23 Derek McNally and Michael Hoskin (1988). William E. Wilson's Observatory at Daramona House. *Journal for the History of Astronomy*, **19**, 146–153. This instrument eventually found a home in the University of London's observatory at Mill Hill, where it gave many years of service.

24 AMC to Gill. 8 December 1893. Cape/RGO.

25 Marilyn Butler (1972). *Maria Edgeworth, A Literary Biography*, Clarendon Press, Oxford. Maria Edgeworth's special interest in science and scientists is recorded in M.T. Brück (1996). Maria Edgeworth: scientific literary lady. *Irish Astronomical Journal*, **23**(1), 49–54.

26 M.T. Brück (1996). The Kuiper–Edgeworth Belt? *Irish Astronomical Journal*, **23**(1), 3; J. McFarland (1996). Kenneth Essex Edgeworth – Victorian polymath and founder of the Kuiper belt? *Vistas in Astronomy*, **40**, 343–51.

27 D. McNally and M. Hoskin, op. cit.

28 W.E. Wilson (1900). *Astronomical and Physical Researches made at Mr Wilson's Observatory, Daramona, Westmeath*. Privately printed.

29 D. McNally and M. Hoskin, op. cit.

30 AMC to W.E. Wilson, 18 June 1900. Letter in the possession of Mr J. Wilson.

31 A.M. Clerke (1895). *The Herschels and Modern Astronomy*. London: Cassell and Co Ltd.

32 The visit is mentioned in the Gill correspondence, Gill to AMC. 16 January 1889. Cape/RGO.

33 *Observatory*, November 1895.

34 J.K. Laughton (1898). *Memoirs of the Life and Correspondence of Henry Reeve CB, DCL*, vol. 2, p. 398. London: Longman.

35 R.A. Marriott (1991). Norway 1896: the BAA's first organised eclipse expedition. *Journal of the British Astronomical Association*, **101**, 162–70.

36 Lady Huggins. *Appreciation*, p. 21.

37 Sir George Baden-Powell. *DNB*

38 [AMC] (1897). Nansen and the Pole. *Edinburgh Review*, **186**, 307–30.

39 Obituary of Shackleton (1871–1921) (1921). *Observatory*, **34**, 255.

40 AMC to Gill. 4 March 1897. Cape/RGO.

41 *Problems in Astrophysics*, pp. 19 and 45.

42 T. Mary Lockyer and Winifred I. Lockyer (1928). *Life and Works of Sir Norman Lockyer*, p. 165. London: Macmillan.

43 C.A. Young (1897). *The Sun*, 4th edition. London: Kegan Paul Trench & Co.

44 D.E. Osterbrock (1894). *Journal for the History of Astronomy*, **15**, p. 81.

45 J.L.E. Dreyer and H.H. Turner (1923), *History of the Royal Astronomical Society*, vol. 1, p. 239. London: Royal Astronomical Society.

46 AMC. *Knowledge*, 1 February 1897.

47 The Royal Observatory Edinburgh was built by the Government at the cost of £30,000 and equipped with first-class instruments from Dun Echt Observatory donated by the Earl of Crawford. A long report, with a photograph, had appeared in *Nature*. Further, the Director was Ralph Copeland, now Astronomer Royal for Scotland, one of Agnes Clerke's oldest friends.

48 E. Walter Maunder (1900). *The Royal Observatory Greenwich, A Glance at its History and Work*. London: The Religious Tract Society.

49 Marilyn Bailey Ogilvie (2000). Obligatory amateurs: Annie Maunder and British women astronomers at the dawn of professional astronomy. *British Journal for the History of Science*, **33**, 67–84.

50 W.H. McCrea (1975). *The Royal Observatory Greenwich*. London: Her Majesty's Stationery Office. A biography of W.M. Christie has yet to be written.

51 *Observatory*, **20**, 143, 1897.

52 AMC (1887). Le progrès del'astronomie depuis soixante ans. *Ciel et Terre*, **18**, 107–15, 273–6.

53 AMC to Keeler. 8 September 1899. MLS.

CHAPTER 10. THE OPINION MOULDER

1 Brian Warner (1979). *Astronomers at the Royal Observatory Cape of Good Hope*, Chapter 5. Cape Town and Rotterdam: A.A. Balkema.

2 AMC to Gill. 16 November 1893. Cape/RGO.

3 AMC to Gill. 8 December 1893. Cape/RGO.

4 Gill to AMC. 22 July 1894, quoted in GF.

5 AMC to Gill, 9 September 1894. Cape/RGO.

6 AMC (1894). *Observatory*, **17**, 234–6.

7 B. Warner. op. cit., p. 98.

8 McClean to Gill. 10 August 1894, quoted in GF, p. 225 and in I.S. Glass, p. 156 (reference below).

9 I.S. Glass (1997). *Victorian Telescope Makers*, p. 155. Bristol and Philadelphia: Institute of Physics.

10 Gill's letters of 11 September 1894 to McClean and to the latter two people are reproduced in GF, pp. 225, 226 and 390.

11 AMC to Gill. 3 October 1894, Cape/RGO.

12 AMC to Gill. 15 July 1898. Cape/RGO.

13 AMC to Gill. 4 October 1894. Cape/RGO.

14 I.S. Glass. op. cit., Chapter 8.

15 B. Warner. op. cit., p. 100. The objective lens had to be returned to Grubb's workshop in Dublin for improvement. The telescope was finally ready only in 1901.

16 Lady Huggins. *Appreciation*, p. 6.

17 AMC to Gill. 9 September 1894. Cape/RGO.

18 AMC to Holden. 21 November 1892. MLS.

19 These and other papers on spectroscopy at that time are discussed by J.B. Hearnshaw (1986). *The Analysis of Starlight*. Cambridge: Cambridge University Press.

20 AMC to Campbell. 28 November 1892. MLS.

21 AMC to Holden. 7 May 1893. MLS.

22 AMC to Campbell. 25 June 1895. MLS.

23 AMC (1895). Some anomalous sidereal spectra. *Observatory*, **18**, 193–5.

24 AMC to Schaeberle. 12 March 1891. MLS.

25 *System of the Stars*, 1st edition, p. 248; 2nd edition, p. 258.

26 AMC to Schaeberle. 10 May 1893. MLS.

27 AMC to Holden. 1 December 1893. MLS.

28 D.E. Osterbrock (1984). The rise and fall of Edward Holden. *Journal for the History of Astronomy*, **15**, 80–127 and 151–76.

29 Ellen Clerke to Holden. 11 December 1892. MLS.

30 AMC to Holden. 18 June 1892. MLS.

31 D.E. Osterbrock. op. cit., p. 108.

32 AMC to Holden. 12 June 1891, MLS.

33 AMC to Holden. 25 February 1894 and 16 March 1894. MLS.

34 AMC to Holden. 5 October 1895. MLS.

35 AMC to Holden. 28 April, 5 May and 12 July 1897. MLS.

36 AMC to Holden. 12 July 1897. MLS.

37 Manuscript Library, United States Military Academy.

38 AMC (1897). *Observatory*, **20**, 52–5.

39 AMC to Pickering. 4 January 1897. HCO.

40 AMC's letter is merely marked 'acknowledged' with the date 18 February 1897.

41 AMC to Pickering. 20 February 1897. HCO.

42 AMC to Pickering. 3 September 1897. HCO. Lady Margaret Domvile
    (1840–1929) contributed to literary journals including the *Cornhill
    Magazine* during the time that Ellen Clerke was writing for it.

43 Lady Domvile's note (undated) to Professor Pickering from Parkes Hotel
    Boston. HCO.

44 AMC to Schaeberle. 22 November 1897. MLS.

45 D.E. Osterbrock (1984). *James E. Keeler: Pioneer American Astrophysicist.*
    Cambridge: Cambridge University Press.

46 ibid., p. 170.

47 AMC to Keeler. 16 June 1899 and 8 September 1899. MLS.

48 AMC to Holden. 15 August 1889. MLS.

49 Holden to AMC. 10 September 1889. MLS.

50 AMC to Holden. 29 September 1889. MSL.

51 Keeler to AMC. 1 November 1890, cited by D.E. Osterbrock. op. cit., p. 99.

52 D.E. Osterbrock. op. cit., p. 315.

53 Keeler to AMC. 29 January 1891, cited by D.E. Osterbrock. op. cit., p. 111.

54 AMC to Keeler. 26 October 1894, cited by D.E. Osterbrock. op. cit., p. 135.

55 AMC to Keeler. 16 June, 1899. MLS.

56 Keeler to AMC. 21 August 1899, MLS.

57 AMC. *The System of the Stars*, Chapter 17 (first edition), p. 265.

58 AMC to Keeler. 8 September 1899. MLS.

59 Campbell to AMC. 28 August 1900. MLS.

60 AMC to Campbell. 25 September 1900. MLS.

61 AMC to Gill. 5 May 1898. Cape/RGO.

62 AMC to Campbell. 12 January 1901. MLS.

63 AMC to Campbell. 25 January 1900. MLS.

64 *Problems in Astrophysics*, p. 224.

65 AMC to Campbell. 28 February 1901. MLS.

66 AMC (1901). Spectrum of Nova Persei. *Observatory*, **24**, 335–8.

[67] AMC to Pickering. 7 April 1901. HCO.

[68] AMC to Gill. 29 June 1902. Cape/RGO.

[69] AMC to Gill. 17 June 1898. Cape/RGO.

[70] AMC to Gill. 9 May 1899. Cape/RGO.

[71] AMC to Gill. 5 July 1901. Cape/RGO.

[72] The lines are from Tennyson.

CHAPTER 11. POPULARISATION, CRYOGENICS AND EVOLUTION

[1] AMC, A. Fowler and J.E. Gore (1898). *Astronomy*. London: Hutchison.

[2] Review in *Nature*, **57**, 266, 1898.

[3] Patrick A. Wayman (1987). *Dunsink Observatory 1785–1985*, p. 124. Dublin: Royal Dublin Society.

[4] Hector MacPherson (Junior) (1905). *Astronomers of Today*, p. 107. London and Edinburgh: Gall and Inglis.

[5] ibid., p. 155.

[6] Reviews of Gore's *Studies in Astronomy* and Ball's *Popular Guide to the Heavens* (1905). *Nature*, **71**, 199.

[7] Camille Flammarion (1911). *Memoires d'un astronome*. Paris: Ernest Flammarion.

[8] Charles C. Gillespie (ed.) (1970–1980). *Dictionary of Scientific Biography*. New York: Scribner's.

[9] Camille Flammarion (translated by J. Ellard Gore) (1897). *Popular Astronomy*. London: Chatto and Windus. Proof of the book's enduring popularity was the publication of a new edition in 1955 prepared by Flammarion's widow, Gabrielle Camille Flammarion, secretary of the French Astronomical Society, and André Danjon, Director of the Paris Observatory. Paris: Flammarion.

[10] Sir William Huggins KCB and Lady Huggins (1899). *An Atlas of Representative Stellar Spectra*. London: William Wesley and Son.

[11] AMC (1900). Representative stellar spectra. *Observatory*, **22**, 308–11.

[12] AMC. *A Popular History of Astronomy during the Nineteenth Century*, 3rd edition, pp. 224–6.

[13] Barbara J. Becker (2000). Priority, persuasion, and the virtue of perseverence: William Huggins' efforts to photograph the solar corona without an eclipse. *Journal for the History of Astronomy*, **31**, 223–43. The problem was solved by Bernard Lyot with his coronagraph in 1930.

[14] AMC to Gill. 5 May 1898. Cape/RGO.

[15] AMC to Gill. 1 June 1899. Cape/RGO.

[16] AMC (1901). *Proceedings of the Royal Institution*, **16**, 699–718.

[17] *DNB*.

18 Gillian Fenwick (1989). *Contributors' Index to the DNB 1885–1901*. Detroit. The scientists among them are listed, with brief notes, in: M.T. Brück (1997). *Irish Astronomical Journal*, **24** (2), 193–8.

19 Footnote to the list of contributors, *DNB*.

20 A.R. Wallace (1903). *Man's Place in the Universe*. London: Chapman and Hall.

21 F.J. Tipler (1981). A brief history of the extra-terrestrial concept. *Quarterly Journal of the Royal Astronomical Society*, **22**, 133–45.

22 AMC to Wallace. 15 March 1901. Add. 46437 British Library.

23 AMC to Wallace. 4 March 1903. Add. 46437 British Library.

24 AMC to Wallace. 17 April 1903. Add. 46437 British Library.

25 *System of the Stars*, 2nd edition, p. 366.

26 AMC to Wallace. 29 April 1903. Add. 46437 British Library.

27 ibid.

28 E.W.M[aunder] (1903). *Knowledge*, **26**, 81–3.

29 AMC to Wallace. 17 April 1903. Add. 46437 British Library.

30 AMC (1903). *Knowledge*, **26**, 108.

31 C. Flammarion (1903). *Knowledge*, **26**, 121.

32 AMC to Wallace. 22 October 1903. Add. 46437 British Library.

33 [AMC] (July 1904). Life in the Universe. *Edinburgh Review*, **200**, 59–74.

34 [AMC] (1896). New views on Mars. *Edinburgh Review*, **184**, 368–85.

35 M.J. Crowe (1986). *The Extra Terrestrial Life Debate 1750–1900*. Cambridge: Cambridge University Press.

36 Deasy family papers.

37 Deasy's original field notebooks and his photographs are preserved among the family papers.

38 H.H.P. Deasy (1901). *In Tibet and Chinese Turkestan*. London: Fisher and Unwin.

39 T. Nigel Brown (1971). *The History of the Manchester Geographical Society 1884–1950*, p. 47. Manchester: Manchester University Press.

40 Deasy family papers.

41 T.H.H[olland?] (1901). *Nature*, **64**, 653.

42 The Deasy Motor Manufacturing Company produced two to three cars a week. In 1912 it became the Siddeley–Deasy Company, until 1919. After Deasy retired from the business this became the famous Armstrong Siddeley Company.

CHAPTER 12. *PROBLEMS IN ASTROPHYSICS*

1 AMC to Gill. 9 September 1894. Cape/RGO.

2 Gill to AMC. 8 March 1898. Cape/RGO.

3 AMC to Gill. Easter Sunday 1898. Cape/RGO.

4   AMC to Gill. 28 June 1898. Cape/RGO.
5   AMC to Gill. 9 May 1899. Cape/RGO.
6   AMC to Gill. 18 February 1900. Cape/RGO.
7   AMC to Gill. 20 June 1901. Cape/RGO.
8   Gill to AMC. 26 December 1901. Cape/RGO.
9   AMC to Gill. 29 June 1902. Cape/RGO.
10 Lady Huggins. *An Appreciation*, p. 20.
11 AMC to Holden. 12 December 1899. MLS.
12 Bacon was the subject of one of Agnes Clerke's *Edinburgh Review* essays in 1879.
13 *History of Astronomy during the Nineteenth Century*, 2nd edition (1893), p. 215.
14 *System of the Stars*, 1st edition (1890), p. 44.
15 A.J. Meadowes (1972). *Science and Controversy, A Biography of Sir Norman Lockyer*, p. 207. Cambridge, Mass: MIT Press.
16 AMC (1896). Five short-period variables. *Observatory*, **19**, 115–16.
17 A. Ritter discussed pulsating stars as a purely theoretical problem in the 1880s (A. Unsöld transl. R.C. Smith (1977). *The New Cosmos*, 2nd edition, p. 210. New York: Springer-Verlag) but the pulsation theory was first seriously applied to stars by H. Shapley in 1914. AMC has a reference to Ritter in her book in connection with stellar temperatures but does not appear to have been aware of his work on pulsating stars.
18 *Problems in Astrophysics*, p. 346.
19 AMC to Gill. 3 December 1903. Cape/RGO.
20 I. Kant (transl. S.J. Jaki) (1981). *Universal History and Theory of the Heavens*, Part 7. Edinburgh: Scottish Academic Press.
21 AMC to Campbell. 7 February 1902. MLS.
22 AMC to Campbell. 17 September 1903. MLS.
23 Gill to AMC. 18 February 1903. GF.
24 Hale to Campbell. 1 April 1903. MLS.
25 AMC to Campbell. 12 April 1903. MLS.
26 H.P. Hollis (1903). *Observatory*, **26**, 170.
27 *Nature* (1903), **68**, 338. Gregory's biography has been written by W.H.G. Armitage (1957). *Sir Richard Gregory, His Life and Work*. London: Macmillan.
28 AMC to Gill. 18 February 1900. Cape/RGO.
29 *Edinburgh Review*, **198**, 122, 1903.
30 *The Academy*, **64**, 173, February 1903.
31 AMC. *DNB*.
32 Obituary of Miss Agnes Mary Clerke. *Times*, 21 January 1907.
33 Quoted in the obituary of Lady Huggins. *Observatory*, **76**, 278, 1916.

CHAPTER 13. WOMEN IN ASTRONOMY IN BRITAIN IN AGNES CLERKE'S
TIME

1   P. Stroobant, J. Devosal, H. Philippot, E. Delporte and E. Merlin (1907). *Les
    Observatoires et les Astronomes*. Brussels: Observatoire Royal de
    Bruxelles. The list updated earlier ones, the first compiled by Edward
    Holden in 1879.

2   Bessie Zapan Jones and Lyle Gifford Boyd (1971). *The Harvard College
    Observatory, The First Four Directorships 1839–1919*, Chapter 11, A field
    for women. Cambridge, Mass.: Harvard University Press; Pamela E. Mack
    (1990). Strategies and compromises: Women in astronomy at Harvard
    College observatory, 1870–1920. *Journal for the History of Astronomy*, 21,
    64–75; Cecilia Payne-Gaposchkin, ed. Katherine Haramundanis (1984). *An
    Autobiography and other Recollections*. Cambridge: Cambridge University
    Press.

3   The one non-British woman is Madame Ceraski of the Astronomical
    Observatory in Moscow, a variable star searcher on photographic plates
    without an official post.

4   Allan Chapman (1999). *The Victorian Amateur Astronomer. Independent
    Astronomical Research in Britain 1820–1920*. Chicester: Praxis Publishing.
    This survey devotes a chapter to these unsung heroes.

5   Chapman. ibid.

6   Phebe Mitchell Kendall (1896). *Life, Letters and Journals of Maria Mitchell*. p.
    111. New York: Houghton Mifflin.

7   W.H. Smyth (1851). *Aedes Hartwelliana*. London (privately printed).

8   Chapman (1999). op. cit.

9   Allan Chapman (1994). The Victorian Amateur Astronomer: William Lassell,
    John Leech, and their worlds. In Patrick Moore (ed.). *1994 Yearbook of
    Astronomy*, p. 163. London: Macmillan.

10 Peter D. Hingley (2001). The first photographic eclipse? *Astronomy and
    Geophysics*, 42, 18–22.

11 A.J. Meadowes (1972). *Science and Controversy, A Biography of Sir Norman
    Lockyer*, p. 67. Cambridge Mass.: MIT Press.

12 H.A. and M.T. Brück (1988). *The Peripatetic Astronomer, The Life of Charles
    Piazzi Smyth*. Bristol: Hilger.

13 'Lectures and Conversations on Geology and Mineralogy by Mr Alexander
    Rose, Fellow of the Royal Scottish Society of Arts, Hon. member of the Jena
    Geological Society etc' (printed syllabus); J. Duncan's geological notes.
    Piazzi Smyth Collection, Royal Observatory Edinburgh.

14 C. Piazzi Smyth (1858). *Report on the Teneriffe Astronomical Experiment of
    1856*. London: Taylor and Francis.

[15] Brück and Brück. op. cit., p. 244.

[16] A. Auwers to Gill. 7 January 1879. GF, p. 97.

[17] I. Gill (1878). *Six Months in Ascension: An Unscientific Account of a Scientific Expedition*. London: John Murray.

[18] J.C. Kapteyn (1914). Sir David Gill. *Astrophysical Journal*, **50**, 161–72.

[19] Allan Chapman (1998). Women in astronomy 1780–1940: summary of an RAS Special Discussion Meeting. *Observatory*, **118**, 270. The wives were those of Sir John Herschel, Sir George Airy and Professor Challis of Cambridge.

[20] M.T. Brück (1998). Lady computers. *Astronomy Now*, **12**(1), 48–51.

[21] M.T. Brück (1995). Lady Computers at Greenwich in the early 1890s. *Quarterly Journal of the Royal Astronomical Society*, **36**, 83–95.

[22] M.T. Brück, op. cit. As some potential applicants were unwilling to work the long hours on such low pay, and the gloss had worn off the scheme, it seemed reasonable to assume that this was the cause. Copeland's letter (below) shows that the real culprit was the Civil Service regulations.

[23] R. Copeland to A. Everett. 14 August 1895. 'It is very unfortunate that after so many years' work at Greenwich ways and means could not be found of retaining you on their permanent staff'. Royal Observatory Edinburgh archives.

[24] Paula Gould (1997). Women and culture in late nineteenth-century Cambridge. *British Journal for the History of Science*, **30**, 127–49. Seven women have been identified and their varied careers followed.

[25] Girton College Register 1869–1946 (1948). Cambridge: Privately printed for Girton College.

[26] M.T. Brück (1994). Alice Everett and Annie Russell Maunder, torch-bearing women astronomers. *Irish Astronomical Journal*, **21**, 281–90.

[27] Gudrun Wolfschmidt (1998). Women astronomers in Germany and America *c.* 1900. In Allan Chapman, Women in Astronomy, an historical perspective, 1780–1940. *Observatory*, **118**, 270–3.

[28] H.C. Vogel (1898). Report of the Astrophysical Observatory Potsdam. *Observatory*, **21**, 279.

[29] M.W. Whitney and A. Everett (1900). *Astrophysical Journal*, **20**, 47 and 76.

[30] Marilyn Bailey Ogilvie (2000). *British Journal for the History of Science*, **33**, 67–84.

[31] AMC (1903). *Problems in Astrophysics*, p. 134.

[32] Annie S.D. Maunder and E. Walter Maunder (1910). *The Heavens and their Story*. London: Charles H. Kelly. The book, written chiefly by Annie, contains her own interesting interpretation of the coronal streamer.

[33] A.S.D. Maunder (1909). Appendix to *Greenwich Observations*. Edinburgh: HM Stationery Office.

[34] M.T. Brück (1998). Lady computers. *Astronomy Now*, 12(1), 48–51. Bellamy is not named in the Brussels list.

[35] Chapman (1999). op. cit.

[36] Chapman (op. cit.) devotes a chapter of his book to these enthusiastic and competent women.

[37] For example, John Evershed's spectroscopy of the flash spectrum at the eclipse of 1898, made before he obtained a professional appointment as Director of Kodaikanal Observatory. (AMC (1902). *History of Astronomy during the Nineteenth Century*, 4th edition, pp. 189 and 200.)

[38] Peggy Aldrich Kidwell (1984). Women astronomers in Britain 1780–1930. *Isis*, 75, 534–46.

[39] Ogilvie. op. cit. Other women, including Agnes Clerke and Margaret Huggins are also, mistakenly, dubbed 'obligatory amateurs'. Agnes Clerke refused the offer of a computership and Margaret Huggins would certainly not have exchanged her superb opportunity to work with the Huggins instruments for a computer's grind.

[40] Chapman distinguishes between Grand Amateurs who could afford to devote themselves fully to astronomy and leisured amateurs whose astronomical pursuits were combined with other activities.

[41] Mary Ward (1858). *The Telescope*. London: Groombridge and sons.

[42] Amédée Guillemin (transl. Mrs Norman Lockyer and edited with additional notes by J.N. Lockyer FRS) (1872). *The Forces of Nature. A Popular Introduction to the Study of Physical Phenomena*. London: Macmillan.

[43] Mary R.S. Creese (1998). *Ladies in the Laboratory? American and British Women in Science 1800–1900: A Survey of their Contributions to Research*, pp. 137–8. Lanham Md. and London: The Scarecrow Press.

[44] M.T. Brück (1998). Mary Acworth Evershed (née Orr) (1867–1949), solar physicist and Dante scholar. *Journal of Astronomical History and Heritage*, 1, 45–59.

[45] M.A. Evershed (1943). Centenary of Agnes Clerke's birth. *Journal of the British Astronomical Association*, 53, 43.

[46] AMC (1898), Among my books, *Literature* (13 August number). The book chosen and described in a light-hearted manner was A. Rebière (1897). *Les Femmes dans les sciences*. Paris.

[47] Chapman (1999). op. cit., p. 286.

CHAPTER 14. REVISED *SYSTEM OF THE STARS*

[1] AMC to Gill. 13 September 1903. Cape/RGO.

[2] AMC to Campbell. 17 September 1903. MLS.

[3] Campbell to AMC. 27 October 1903, mentioned in the later correspondence. MLS.

[4]   AMC to Pickering. 5 November 1903. HCO.

[5]   AMC to Pickering. 1 January 1904. HCO.

[6]   AMC. *Problems in Astrophysics*, p. 120.

[7]   AMC to Hale. 26 June 1903. Yerkes.

[8]   Hale to AMC. 27 May 1903. Cal Tech.

[9]   Helen Wright (1994). *Explorer of the Universe, A Biography of George Ellery Hale*, p. 149. New York: American Institute of Physics.

[10] Hale to AMC. 27 May 1903. Cal Tech.

[11] AMC to Hale. 26 June 1903. Cal Tech.

[12] AMC to Hale. 13 December 1903. Cal Tech.

[13] Hale to AMC. 4 February 1904. Cal Tech.

[14] Hale to AMC. ibid.

[15] Hale to AMC. 7 March 1904. Cal Tech.

[16] AMC to Hale. 19 February 1904. Cal Tech.

[17] AMC to Hale. 26 March 1904. Cal Tech.

[18] AMC to Hale. 23 March 1904. Cal Tech.

[19] Hale to AMC. 12 April 1904. Cal Tech.

[20] AMC to Hale. 19 February 1904. Cal Tech.

[21] Simon Newcomb (1902; originally published 1901). *The Stars, A Study of the Universe*. London: John Murray.

[22] AMC to Gill. 21 November 1901. Cape/RGO.

[23] *Problems in Astrophysics*, p. 14.

[24] AMC. *History of Astronomy during the Nineteenth Century*, 3rd edition, p. 379; 4th edition, p. 312.

[25] D.E. Osterbrock (1994). *James Keeler Pioneer American Astrophysicist*, p. 317. Cambridge: Cambridge University Press.

[26] AMC. *System of the Stars*, 2nd edition, p. 373.

[27] AMC. *System of the Stars*, 2nd edition, p. 260.

[28] AMC. *System of the Stars*, 2nd edition, p. 280.

[29] AMC to Campbell. 5 January 1904. MLS.

[30] AMC to Campbell. 24 March 1904. MLS.

[31] Campbell to AMC. 8 April 1904. MLS.

[32] AMC to Campbell. 28 April 1904. MLS.

[33] Gill to AMC. 1 November 1904. Cape/RGO. The plate was dispatched on that date.

[34] AMC to Gill. 8 December 1904. Cape/RGO.

[35] Campbell to AMC. 8 December 1905. MSL.

[36] *Nature*, **73**, 505, March 1906.

[37] AMC. *System of the Stars*, 2nd edition, p. 76.

[38] AMC. *System of the Stars*, 2nd edition, p. 44.

39 B. Lightman (1997). Constructing Victorian Heavens, Agnes Clerke and the 'New Astronomy'. In B.T. Gates and A.B. Shteir (eds.). *Natural Eloquence, Women Reinscribe Science*. Madison: University of Wisconsin Press.

40 R.A.G[regory] (1906). *Nature*, **74**, 350.

41 AMC (1905). *Modern Cosmogonies*. London: A.and C. Black.

42 [AMC] (1898). Ethereal telegraphy. *Edinburgh Review*, **188**, 297–310.

43 [AMC] (1903). The revelations of radium. *Edinburgh Review*, **198**, 374–99.

44 Norman Feather (1973). *Lord Rutherford*, p. 93. London: Priory Press Ltd. The two quotations are from a letter to Rutherford from Larmor, Secretary of the Royal Society.

45 Eve Curie (transl. V. Sheean) (1942). *Madame Curie*, p. 201. London: The Reprint Society (originally published by Heinemann, London, 1938).

46 ibid.

47 Sir William Huggins KCB and Lady Huggins (1903). *Proceedings of the Royal Society*, **72**, 196–9 and 409–13.

48 'There were still some dissentients [after the publication of Rutherford and Soddy's disintegration theory in May 1903] but in spite of them, the disintegration theory was "on a flood-tide of interest".' (N. Feather, op. cit. p. 88)

49 AMC to Gill. 13 September 1903. Cape/RGO.

50 Susan Quinn (1996). *Marie Curie, A Life*, p. 185. London: Mandarin Paperbacks.

51 AMC (1907). Old and New Alchemy. *Edinburgh Review*, **205**, 28–47.

52 E.M. Clerke (1901). *Dublin Review*, **123**, 227–55.

53 The Cardinal was not without his critics. *The Glasgow Observer and Catholic Herald* strongly disagreed with his imperial views in an editorial on 6 January 1900 (reprinted in the *Scottish Catholic Observer*, 1 January 2000).

54 Obituary. *The Tablet*, 10 March 1906.

55 E.M. Clerke (1881). *The Flying Dutchman and Other Poems*. London: Satchell and Co.

56 E.M. Clerke (1899). *Fable and Song in Italy*. London: Grant Richardson.

57 Review. *Literature*, 12 August 1899, 139–40.

58 R. Garnett (1898). *A History of Italian Literature*. London: Heinemann.

59 E.M. Clerke (1902). *Flowers of Fire*. London: Hutchison and Co.

60 Obituary notice in the *Tablet*, op. cit. I have not traced the Italian journal in question.

CHAPTER 15. COSMOGONIES, COSMOLOGY AND NATURE'S SPIRITUAL CLUES

1 Lady Huggins. *Appreciation*, p. 30.

2 *The System of the Stars*, pp. ix and 11.

3   Mary Somerville (1835). *The Connexion of the Physical Sciences*, p. 2. London: John Murray.

4   In Sir John Herschel (1857). *Essays*. London: Longman.

5   Bernard Lightman (1997). The voices of nature: popularising Victorian science. In B. Lightman (ed.). *Victorian Science in Context*, pp. 202–4. Chicago: University of Chicago Press.

6   Bernard Lightman (1997). Constructing Victorian Heavens. In Barbara T. Gates and Ann B. Shteir (eds.). *Natural Eloquence, Women Reinscribe Science*, p. 73. Madison: University of Wisconsin Press.

7   *Aeterni Patris* (1879).

8   Edward Duffy (1997). *Saints and Sinners, A History of the Popes*, p. 241. Yale: Yale University Press.

9   *Panis Angelicus* comes from the hymn *Sacris Solemnis*. The others are *Pange Lingua, Verbum Supernum* and the sequence *Lauda Sion*.

10 Michael Hoskin and Owen Gingerich (1997). Medieval Latin Astronomy. Chapter 4 in Michael Hoskin (ed.) *The Cambridge Illustrated History of Astronomy*, p. 75. Cambridge: Cambridge University Press.

11 J.L.E. Dreyer (1906). *A History of Astronomy from Thales to Kepler*, formerly titled *History of the Planetary Systems from Thales to Kepler* (1906). New York: Dover Publications (1953).

12 Dreyer. op. cit., p. 232.

13 F.C. Copleston (1955). *Aquinas*, p. 13. London: Penguin (The Pelican Philosophy Series).

14 The 'five ways' are discussed by Copleston, op. cit.

15 S.J. Jaki (1989). *God and the Cosmologists*. Edinburgh: Scottish Academic Press; (1980). *Cosmos and Creator*. Edinburgh: Scottish Academic Press.

16 H.A. Brück (1968). *Theology and the Physical World*, Lecture to the Theological Club of New College, Edinburgh, delivered 12 March. (unpublished).

17 Avery Dulles S.J. (1990). Science and theology. In R.J. Russell, W.R. Stoeger and G.V. Coyne (eds.). *John Paul II on science and religion*, pp. 13–14. Vatican Observatory Publications (distributed by the University of Notre Dame Press, Notre Dame, Indiana).

18 AMC (1901). The Hodgkins Trust Essay: Low temperature research at the Royal Institution, *Proceedings of the Royal Institution*, **16**, 699–719.

19 J.R. Gasquet (1885). Arguments for the existence of God. *Dublin Review*, **14**, 65–78. The author was the physician brother of the English Cardinal Gasquet, historian and Benedictine Abbot.

20 Joseph Rickaby S.J. (1889). What has the Church to do with Science? *Dublin Review*, **14**, 243–53.

[21] John H. Vaughan (1889). Faith and Reason. *Dublin Review*, **22**, 72–82.

[22] *System of the Stars*, 1st edition, p. 381.

[23] *System of the Stars*, 1st edition, p. 380; 2nd edition, p. 361.

[24] *System of the Stars*, 1st edition, p. 82. This and the reference that follows occur in a chapter on sidereal evolution, dropped from the second edition.

[25] *System of the Stars*, 1st edition, pp. 82 and 84.

[26] St Thomas Aquinas. *Summa Contra Gentiles*, quoted by Brück, op. cit.

[27] She discussed Kant at length in *Modern Cosmogonies*. Her famous article on Laplace in *Encyclopaedia Britannica* had marked the beginning of her career as a serious interpreter of science.

[28] *System of the Stars*, 1st edition, p. 94.

[29] *Problems in Astrophysics*, Introduction, p. 10.

[30] E.W.M[aunder] (1903). *Knowledge*, **26**, 81–83. It did not appear so to everybody. In the debate with A.R. Wallace about Man in the Universe (Chapter 11) E.W. Maunder, an active evangelical Christian, seemed to maintain the opposite. Wallace's underlying error, he said, was that 'he has reasoned from the area which we can embrace with our limited perception to the Infinite beyond our mental and intellectual grasp. We are on the earth, and can only reason, only guess, from our earthly experience of the laws, of the materials, of the conditions, elsewhere'.

[31] *Problems in Astrophysics*, p. 10.

[32] The extragalactic expanding universe that eventually emerged was to pose no threat to St Thomas.

[33] *System of the Stars*, 2nd edition, p. 349.

[34] Simon Newcomb (1902). *The Stars, A Study of the Universe*, p. 319. London: John Murray.

[35] *Modern Cosmogonies*, p. 249.

[36] *System of the Stars*. These are the final words in both editions of the book.

[37] '. . . the sky vanished, as a scroll rolled up.' The Apocalypse or Revelation of John, Chapter 6, Verse 14. (*The New English Bible*, 1970. Oxford University Press, Cambridge University Press.)

[38] J.P. Nichol (1839). *The Architecture of the Heavens*, 3rd edition, p. 195. Edinburgh: William Tait.

[39] Bernard Lightman (1989). Ideology, evolution and late-Victorian agnostic popularisers. In James R. Moore (ed.). *History, Humanity and Evolution*. Cambridge: Cambridge University Press.

[40] Nevertheless, Spencer's essay on the Nebular Hypothesis (1854) was one of the authorities quoted by Agnes Clerke in support of the one-system universe. It is hard to understand today why Spencer was regarded as such an expert on the subject.

CHAPTER 16. LAST DAYS AND RETROSPECT

[1] D.E. Osterbrock (1964). *James Keeler Pioneer American Astrophysicist*, p. 153. Cambridge: Cambridge University Press.

[2] Campbell to AMC. 8 April 1904. MLS.

[3] AMC to Campbell. 28 April 1904. MLS.

[4] George Ellery Hale (1932). *Signals from the Stars*, pp. 51–2. London: Charles Scribner.

[5] *History of Astronomy during the Nineteenth Century*, 3rd edition, p. 246.

[6] *History of Astronomy during the Nineteenth Century*, 4th edition, p. 198.

[7] *Problems in Astrophysics*, p. 98.

[8] *System of the Stars*, 2nd edition, p. 42.

[9] Margaret Huggins to Hale. 4 October 1905. Cal Tech.

[10] Margaret Huggins to Hale. 30 January 1907. Cal Tech. This same letter conveyed the news of Agnes Clerke's death.

[11] AMC to Hale. 12 August 1906. Cal Tech.

[12] Hugh Chisholm was the brother of the mathematician Grace Chisholm Young (Chapter 13).

[13] AMC to Newcomb. 27 May 1898. Library of Congress.

[14] M.T. Brück (1997). Agnes Clerke's work as a scientific biographer. *Irish Astronomical Journal*, **24(2)**, 193–7.

[15] AMC to Newcomb. 14 March 1906. Library of Congress.

[16] AMC to Newcomb. 23 April 1906. Library of Congress.

[17] [AMC] (1906). A representative philosopher. *Edinburgh Review*, **104**, 157–78.

[18] Gill to AMC. 15 November 1905. Cape/RGO.

[19] Gill to A. Roberts. 11 September 1905. GF, p. 240.

[20] Brian Warner (1979). *Astronomers at the Royal Observatory Cape*. Rotterdam and Cape Town: A.A. Balkema.

[21] AMC to Gill. 3 December 1905. Cape/RGO.

[22] Margaret Huggins to G.E. Hale. 30 January 1907. Cal Tech.

[23] Obituary. *Tablet*, 26 January 1907.

[24] AMC (1907). *Observatory*, **30**, 55.

[25] Newcomb to Gill. 5 March 1907. GF, p. 337.

[26] E.E. Barnard, quoted by T.J.J. See, reference below.

[27] T.J.J. See (1907). Some recollections of Miss Agnes M. Clerke. *Popular Astronomy*, **15**, 323–7.

[28] *Times* (London). 21 January 1907. Other obituary notices and reminiscences appeared in *Monthly Notices of the RAS* (M.L. Huggins); *Popular Astronomy* (T.J.J. See and Hector Macpherson Jr.); *Nature*; *Observatory*; *Knowledge* (T.E. Heath).

[29] *Munster Advertiser*, 26 January 1907.

30 Obituary (1907). *Observatory*, **30**, 107–8.
31 Hector Macpherson Jr. (1907). Miss Agnes Mary Clerke. *Popular Astronomy*, **15**, 165–8.
32 E.S.G. (1907). The Late Miss Agnes M. Clerke. *Knowledge and Scientific News*, **4**, 31–2.
33 Deasy family papers. A list, and a number of clippings.
34 Lady Huggins. *An Appreciation*.
35 Lady Huggins. op. cit.
36 Arthur Berry (1961). *A Short History of Astronomy from Earliest Times Through the Nineteenth Century*. London: Dover (originally published 1898).
37 Kevin Krisciunas (1894). Translator's Preface. In D.B. Hermann. *The History of Astronomy from Herschel to Hertzsprung*. Cambridge: Cambridge University Press.
38 Helen Wright (1994). *Explorer of the Universe, A Biography of George Ellery Hale*. New York: American Institute of Physics (Original edition New York: Dutton 1966).
39 Stanley Jaki (1972). *The Milky Way, An Elusive Road for Science*. New York: Science History Publications.
40 M. Hoskin (ed.) (1984). *The General History of Astronomy*, vol. 4A, Chapters 1, 4, 5, 6 and Figures 2.5 and 9.2. Cambridge: Cambridge University Press.
41 Michael J. Crowe (1994). *Modern Theories of the Universe from Herschel to Hubble*. New York: Dover Publications Ltd.
42 Michael Hoskin (ed.) (1997). *The Cambridge Illustrated History of Astronomy*. Cambridge: Cambridge University Press.
43 Michael Hoskin (1999). Note on Agnes Clerke written for the author.
44 *DNB*.
45 Cecilia Payne-Gaposhkin (1996). *An Autobiography and other Recollections*, 2nd edition, Chapter 2. Cambridge: Cambridge University Press.
46 W.H.W[esley] (1895). *Knowledge*, **18**, 34–6.
47 Lady Huggins. *An Appreciation*.
48 Preface to the last edition (1902) of *A Popular History of Astronomy during the Nineteenth Century*.

CHAPTER 17. EPILOGUE

1 L.E. Anderson and E.A. Whitaker (1982). *NASA Catalogue of Lunar Nomenclature*.
2 *Southern Star*, Skibbereen, 17 July 1999.

Appendix

Some short passages from Agnes Clerke's writings are given which give an idea of her style of writing as well as of her versatility.

Precursors of Newton

(from *Edinburgh Review*, 1879)

The problem of gravity was the supreme question of that time. It stood first among the orders of the day of the scientific council. It was instinctively felt that until it should be disposed of, no real progress could be made in physical knowledge and, slowly, but surely, the way was being prepared for a great discovery. Galileo had made Newton possible. Men's ideas were gradually clarifying; the great cosmical analogies, now so familiar, were, step by step emerging out of the dusk of ignorance; antiquated prepossessions were sinking in a sediment of cloudy cavil, out of sight. Heaven was assimilated to earth, and earth to heaven; the old gratuitous separation between the starry firmament over our heads and the solid globe under our feet was abolished by acclamation; and it was felt that the coming law, to be valid, must embrace in its operation the whole of the visible universe. Towards this consummation Gilbert contributed something by his theory of universal magnetism; and Galileo, as well as Bacon and Horrocks foresaw that in this direction lay the coveted secret.

Solar–terrestrial relations

(from an essay on Aurorae, *Edinburgh Review*, 1886)

In some unknown manner, solar energy unquestionably reacts upon the electrical and magnetic condition of our planet, by turns stimulating and relaxing in correspondence with its own rhythmical alternations. Such correspondence is part of the vital union which subsists between the various members of our system. No one of them is truly independent of the others. They form together, as it were, a living and cooperative whole. Polar lights are as a beacon kindled in response to the subtle messengers of the sun, telling that the Earth is no dead or cast-off member, but still thrills in harmony with, as well as moves obediently to, the great life-sustaining luminary.

Mars

(concluding part of a review of books by Lowell, Flammarion and Schiaparelli, *Edinburgh Review*, 1896)

There is then no compulsion on us to regard the surface of Mars as modelled to suit their vital needs by the industry of rational creatures. Irrigation hypotheses, inland navigation hypotheses and the like are superfluous, and being superfluous, are inadmissible. Not that they are, in all shapes, demonstrably false, but that they open the door to pure license in theorising. The admission of vegetable growth and decay as an element of visible change is less objectionable, and is apparently capable of being justified spectroscopically, but until that some other kind of definite evidence is forthcoming, the subject invites only nebulous conjecture . . .

We venture to disclaim, on behalf of humanity, the extramundane jealousy imputed to it by Mr Lowell. At the close of the nineteenth century, after so many poignant disillusions, and the wreck of so many passionate hopes, it is not enamoured with its own destinies to the point of desiring to impose them as a maximum of happiness on the universe. Rather, men cherish the vision of other and better worlds, where intelligence untrammelled by moral disabilities may have risen to unimaginable heights, and sense and reason alike are dominated by incorrupt will. But it is improbable that the vision can ever be treated in any one of the disseminated orbs around us. The problem of universal life is an enticing, yet insoluble, one.

The unification of physics and astronomy

(from *History of Astronomy in the Nineteenth Century*, pp. 175–6, 1893)
Nearly three centuries ago, Kepler drew a forecast of what he called 'physical astronomy' – a science treating of the efficient causes of planetary motion, and holding the key to the 'inner astronomy'. What Kepler dreamed of and groped after, Newton realised . . . [with the discovery of gravity] . . . The world under our feet was thus for the first time brought into physical connection with the worlds peopling space, and a very tangible relationship was demonstrated as existing between what used to be called the 'corruptible' matter of the earth and the 'incorruptible' matter of the heavens . . .

The establishment of the new method of spectrum analysis drew far closer this alliance between celestial and terrestrial science . . . Up to the middle of the present century, astronomy, while maintaining her strict union with mathematics, looked with indifference on the rest of the sciences; it was enough that she possessed the telescope and the calculus. Now the materials for her inductions are supplied by the chemist, the electrician, the inquirer into the most recondite mysteries of light, and the molecular constitution of matter . . . Her position of lofty isolation has been exchanged for one of community and mutual aid. The astronomer has become, in the highest sense of the term, a physicist; while the physicist is bound to be something of an astronomer.

Accumulating data without purpose

(comment on a certain over-ambitious project) (*System of the Stars*, second edition, pp. 368–9, 1905)
The life of a science is in the thought that binds together the facts; decadence has already set in when they come to be regarded as an end in themselves. 'Man is the interpreter of Nature'; to draw up an inventory, however, is not to interpret. It is true that speculation is prone to wander into devious ways: but then 'truth emerges more easily from error than from confusion'. And in sidereal science especially, there is a danger lest investigators, seduced by the wonderful facilities of novel methods, should exhaust their energies

upon the accumulation of data, and leave none for the higher work of marshalling them along the expanding lines of adequate theory.

The nebular hypothesis varied and improved

(from *Modern Cosmogonies*, Chapter 4, pp. 60–61, 1905. A translation into English of Kant's treatise had recently been published.)

'Restorations' often go very far. Things may be improved beyond recognition, nay, out of existence. So it has happened with the nebular hypothesis. *Stat nominis umbra.* The name survives, but with connotations indefinitely diversified. The original theme is barely recalled by many of the variations played upon it. Entire license of treatment prevails. The strict and simple lines of evolution laid down by Laplace are obliterated or submerged. Some of the schemes proposed by modern cosmogonists are substantially reversions to Kant's *Natural History of the Heavens*; the long-discarded and despised Cartesian vortices reappear, with the eclat of virtual novelty, in others; nor are there wanting theories or speculations reminiscent even of Buffon's cometary impacts. Moreover, the misleading fashion has come into vogue of bracketing Kant with Laplace as co-inventor of the majestic and orderly plan of growth, commonly designated the 'nebular hypothesis'. This has been, and is, the source of much confusion. Save the one fundamental idea – and that by no means their exclusive property – of ascribing unity of origin to the planetary system, Kant's and Laplace's evolutionary methods had little in common. Their postulates were very far from being identical; they employed radically different kinds of 'world-stuff'; and the 'world-stuff' was subjected, in each case, to totally dissimilar processes.

Yet it is often assumed that to defend or refurbish one scheme is to rehabilitate the other. Under cover of the intellectual vagueness thus fostered, a backward drift of thought is, indeed, discernible towards the view-point of the Königsberg philosopher. It is recommended, not so much by the favourable verdict of science as by the wide freedom of the prospect it affords. The imperative guidance of Laplace, reasssuring at first, led to subsequent revolts. But Kant is highly accommodating; one can deviate widely from, without finally quitting, the track of his conceptions; they are capacious and indefinite enough to comport with

much novelty both of imagination and experience, and hence lend
themselves to the changing requirements of progress.

Needlework as Art

(from a review of books on tapestry and needle work, *Edinburgh Review*, 1886)
It has become the fashion to take things seriously. 'Movements' in all
manner of directions, and for all sorts of purposes, are the order of the
day. Their resultant effect forms a curious problem in social dynamics,
of which we, at least, shall not adventure a solution. Meanwhile, we
may thankfully accept whatever good ensues from attempts at reform
prompted by amiable and exaggerated fervours. Enthusiasm is bound to
be one-sided; else it would lack motive power. Limited faculties are apt
to become paralysed by too wide a survey of possibilities.

Our present concern is with a species of border zone, or district
of divided ownership, between two wide provinces of modern
organised endeavour. Efforts towards improving the position of women
cannot be better expended than in giving higher significance to work
especially appropriated to them. Efforts towards raising the standard of
art can scarcely exclude that primitive mode of decoration in which
the needle replaces the pencil and variously tinted threads are
substituted for pigments. To some, whose conception of the attainable
with such materials was fixed by the cross-stitch and tent-stitch
abominations of our grandmothers, and who have ignored the earnest
and not unsuccessful studies of the last score of years, it may, indeed,
even still, be a new idea that needlework is, or may be, a branch of art.
But we venture to assert that these doubts will cease after a perusal,
however cursory, of the beautiful volume which has suggested our
present theme. [i.e. *Needlework as Art*, by Lady Marian Alford, 1886.
Lady Alford was patroness of the Royal School of Art.]

Caroline Herschel

(from *The Herschels and Modern Astronomy*, pp. 139–141, 1895)
Caroline Herschel was not a woman of genius. Her mind was sound
and vigorous rather than brilliant. No abstract enthusiasm inspired

her; no line of inquiry attracted her; she seems to have remained ignorant even of the subsequent fate of her own comets. She prized them as trophies, but not unduly. The assignment of property in comets reminded her, she humorously remarked, when in her ninety-third year, of the children's game, 'He who first cries 'Kick' shall have the apple.' Yet her faculties were of no common order, and they were rendered serviceable by moral strength and absolute devotedness. Her persistence was indomitable, her zeal was tempered by good sense; her endurance, courage, docility, and self-forgetfulness went to the limits of what is possible in human nature. With her readiness of hand and eye, her precision, her rapidity, her prompt obedience to a word or glance, she realised the ideal of what an assistant ought to be . . .

The aim of the life of this admirable woman was not to become learned or famous, but to make herself useful. Her function was, in her own unvarying opinion, a strictly secondary one. She had no ambition. Distinctions came to her unsought and incidentally. She was accordingly content with the slight and fragmentary supply of knowledge sufficing for the accurate performance of her daily tasks. No inner craving tormented her into amplifying it. The following of any such impulse would probably have impaired, rather than improved, her efficiency. The turn of her mind was above all things practical. She used formulae as other women use pins, needles, and scissors, for certain definite purposes, but with complete indifference as to the mode of their manufacture. What was required of her, however, she accomplished superlatively well, and this was the summit of her desires. She shines, and will continue to shine, by the reflected light that she loved.

Night's soliloquy

Poem by Ellen Clerke from *The Flying Dutchman and other Poems*, London 1881 (written when Agnes began writing *A Popular History of Astronomy during the Nineteenth Century*).

> Who calls me dark? for do I not display
>   Wonders that else man's eye would never see?
> Waste in the blank and blinding glare of Day
>   The heavens bud forth their glories but to me.

Is it not mine to pile their crystal cup,
   Drained by the thirsty sun and void by day,
Brimful of living gems, profuse heaped up
   The bounteous largesse of my royal way?

Mine to call o'er at dusk the roll of heav'n
   Array its glittering files in order due?
To beckon forth the lurking star of Even
   And bid the constellations start to view?

The wandering planets to their paths recall
   And summon to the muster tenant spheres,
Till thronging to my standard one and all
   They crowd the zenith in unfathomed tiers?

Do I not lure stray sunbeams from the day,
   To hurl them broadcast as winged meteors forth?
Strew sheaves of fiery arrows on my way
   And blazon my dark spaces in the north?

Is not a crown of lightnings mine to wear,
   When polar flames suffuse my skies with splendour?
And mine the homage with the sun to share,
   His vagrant vassals rush through space to render?

Who calls me secret? are not hidden things,
   Revealed to science when with piercing sight
She looks beneath the shadow of my wings
   To fathom space and sound the infinite?

In plasmic light do I not bid her trace
   Germs from creation's dawn, maturing slow?
And in each filmy chaos drowned in space
   See suns and systems yet in embryo?

Bibliography

A.M.CLERKE'S SCIENTIFIC BOOKS

A Popular History of Astronomy during the Nineteenth Century (1885).
   Edinburgh: Adam and Charles Black. 2nd edition (1887). 3rd edition (1893).
   4th edition (1902).

The System of the Stars (1890). London: Longman. 2nd edition (1905). London: A.
   and C. Black.

Problems in Astrophysics (1902). London: A. and C. Black.

Modern Cosmogonies (1905). London: A. and C. Black.

BIBLIOGRAPHY

Becker, B.J. (1996). Dispelling the myth of the able assistant: William and Margaret
   Huggins at the Tulse Hill observatory. In *Creative Couples in the Sciences*,
   eds. H.M. Pycor *et al.* New Brunswick: Rutgers University Press.

Brück, M.T. (1993). Ellen and Agnes Clerke of Skibbereen, scholars and writers.
   Seanchas Chairbre, **3**, 23–43.

Brück, M.T. (1994). Agnes Mary Clerke, chronicler of astronomy. *Quarterly
   Journal of the Royal Astronomical Society*, **35**, 59–79.

Chapman, Allan (1999). *The Victorian Amateur Astronomer. Independent
   Astronomical Research in Britain 1820–1920.* Chichester: Praxis
   Publishing.

Creese, Mary R.S. (1998). *Ladies in the Laboratory? American and British Women
   in Science 1800–1900.* Lanham, Md. and London: The Scarecrow Press Inc.

Crowe, Michael J. (1994). *Modern Theories of the Universe from Herschel to
   Hubble.* New York: Dover Publications Inc.

Dreyer, J.L.E and Turner, H.H. (eds.) (1987). *History of the Royal Astronomical
   Society, Vol 1, 1820–1920.* Oxford: Blackwell Scientific Publications.
   (originally published 1923).

Duffy, Eamon. (1997). *Saints and Sinners, A History of the Popes.* Yale University
   Press.

Forbes, George (1916). *David Gill, Man and Astronomer.* London: John Murray.

Hearnshaw, J.B. (1986). *The Analysis of Starlight.* Cambridge University Press.

Hoskin, Michael (ed.) (1997). *The Cambridge Illustrated History of Astronomy*. Cambridge University Press.

Jaki, Stanley L. (1980). *Cosmos and Creator*. Edinburgh: Scottish Academic Press.

Lightman, Bernard (1997). 'The voices of Nature'; popularising Victorian science. In *Victorian Science in Context*, ed. B. Lightman. University of Chicago Press.

Macpherson Jnr., Hector (1905). *Astronomers of Today and their Work*. London and Edinburgh: Gall and Inglis.

McCrea, W.H. (1975). *The Royal Greenwich Observatory*. London: Her Majesty's Stationery Office.

Meadowes, A.J. (1972). *Science and Controversy, A Biography of Sir Norman Lockyer*. Cambridge, Mass: MIT Press.

O'Neil, Robert (1955). *Cardinal Herbert Vaughan*. London: Burns and Oates.

Osterbrock, Donald E. (1984). *James E. Keeler: Pioneer American Astrophysicist*. Cambridge University Press.

Osterbrock, Donald E. (1984). The rise and fall of Edward S. Holden. *Journal of the History of Astronomy*, **15**, 81–127 and 151–76.

Warner, Brian (1979). *Astronomers at the Royal Observatory Cape of Good Hope*. Cape Town and Rotterdam: A.A.Balkema.

Wright Helen (1994). *Explorer of the Universe, A Biography of George Ellery Hale*. New York: American Institute of Physics.

Index

Page numbers in italics indicate illustrations.